Rüdiger Wapler

Unemployment, Market Structure and Growth

 Springer

Author

Rüdiger Wapler
University of Tübingen
Faculty of Economics
Mohlstrasse 36
72074 Tübingen
Germany

Cataloging-in-Publication Data applied for

A catalog record for this book is available from the Library of Congress.

Bibliographic information published by Die Deutsche Bibliothek
Die Deutsche Bibliothek lists this publication in the Deutsche Nationalbibliografie;
detailed bibliographic data is available in the Internet at http://dnb.ddb.de

ISSN 0075-8450
ISBN 978-3-540-40449-1   ISBN 978-3-642-55893-1 (eBook)
DOI 10.1007/978-3-642-55893-1

http://www.springer.de

© Springer-Verlag Berlin Heidelberg 2003

Originally published by Springer-Verlag Berlin Heidelberg New York in 2003

Typesetting: Camera ready by author
Cover design: *Erich Kirchner*, Heidelberg

Printed on acid-free paper   55/3143/du   5 4 3 2 1 0

Lecture Notes in Economics and Mathematical Systems

530

Springer-Verlag Berlin Heidelberg GmbH

To Karin

Foreword

In his Ph.D. thesis, Rüdiger Wapler analyses the causes of the persistently high unemployment rates especially in continental Europe. Particular emphasis is placed on imperfect labour and product markets on the one hand, and on the numerous links between unemployment, innovations and growth on the other. Hence, Rüdiger Wapler provides an important contribution towards a better understanding of both the development of labour markets as well as the dynamics of growth.

To aid readers with only little prior knowledge of labour markets, the book presents the most common theories of unemployment: (1) trade-union models in which union bargaining power leads to wages above their market-clearing level, (2) efficiency-wage models in which employers voluntarily pay higher wages in order to motivate or discipline their workers or to reduce the job-turnover rate, as well as (3) matching models in which unemployment is caused by the continuous turnover of jobs and workers. In addition, emphasis is placed on the fact that labour needs to be treated as heterogeneous, a fact often neglected in the literature. Subsequently, these labour-market foundations are integrated with modern theories of innovations and growth, making the approach much more relevant and plausible. Without doubt, the generalisations of the models performed by Rüdiger Wapler show that there are limits to such formal analysis. Due to the increasing number of interdependencies, it is doubtful whether even more complex models provide additional (usable) insights. This book is aimed at economists researching on labour markets, innovations and growth. I hope it receives the attention it deserves.

Professor Dr. Manfred Stadler

Acknowledgements

This book had its origins as a Ph.D. dissertation in economics at the University of Tübingen, Germany. I am deeply grateful to my primary advisor, Manfred Stadler. I was able to greatly benefit from his far-reaching knowledge of economics and excellent teaching skills. Both are invaluable assets. Further, his sense of humour and the priority he placed on creating an enjoyable atmosphere at the department made working here highly enjoyable. I am just as indebted to my colleague and friend Stephan Hornig. The time and effort he took to read my many drafts, his sound judgement and his advice made him simply fun to work and be with. The quality of my work profited enormously from him. Thanks also go to my secondary advisor, Uwe Walz, who was always prepared to listen, advise and help.

Writing a dissertation is often a lonely period fraught with frustrations. Karin Buchenau is undoubtedly the person who had to suffer most from my downs but still held to me, encouraged me and gave me new drive. She made my frustrations tolerable and my successes and accomplishments more joyful. I cannot thank her enough. My many friends and colleagues at the University of Tübingen also deserve a great deal of gratitude for making my experience more enjoyable. In particular, I want to mention Jürgen Henrion, Patrick Herbst, Andreas Scheuerle, Cornelia Neff, Elke Amend, Cornelia Kaldewei, Ralf Münnich, Dirk Baur, Astrid Hellwig, Katharina Hauser, Petra Kopf, Matthias Weiss, Stephan Lengsfeld and Leslie Neubecker.

I also want to specially mention my first co-authors, Axel Schimmelpfennig and Dietmar Hornung. Axel has been an inspiration and motivation from day one of my economic studies and was the perfect study partner. Both deepened my understanding of economics and made arguing about economics enjoyable.

As in all my endeavours, my parents Karin and Horst Wapler were very supportive. I am grateful to my father for having implanted and nourished my intellectual curiosity and to my mother for her moral support.

Tübingen, April 2003   Rüdiger Wapler

Contents

List of Figures

List of Tables

List of Symbols

a	Job-finding rate
a_{sS}	Partial derivative of the unit cost function with respect to wages for labour of type s in sector S
b	Job-breakup rate
b_s	Job-breakup rate for labour of type s
c	Costs of searching for and hiring a worker
$\hat{c}$	Consumption level measured in efficiency units
c_j	Unit costs of producing the intermediate good j
c_S	Unit costs in sector S
d	Distribution factor
$\tilde{e}$	Effort
$\tilde{e}_{sS}$	Effort of type s labour in sector S
$\tilde{e}_s$	Effort of type s labour
$\tilde{e}^*$	Optimal effort level
f	Rate at which vacancies are filled
f_s	Rate at which vacancies are filled for labour of type s
g_A	Growth rate of technological progress
g	Growth rate of final output
g^*	Steady-state growth rate of final output
h_j	Starting index for number of quality improvements of intermediate good j

i	Firm or brand index
$\tilde{i}$	Firm or brand index with $i \neq \tilde{i}$
j	Intermediate goods index
$\tilde{j}$	Component index
$\hat{k}$	Capital intensity measured in efficiency units
$\hat{k}^*$	Steady-state capital intensity measured in efficiency units
l_j	Running index for number of quality improvements of intermediate good j
m_i	Quantity of the manufacturing good of brand i
$\bar{m}_j$	Number of qualitative improvements of intermediate good j
$\bar{m}_{\tilde{j}}$	Number of qualitative improvements of component $\tilde{j}$
$\tilde{m}$	Matching rate
n	Number of varieties of the manufacturing good
$\bar{n}$	Number of different intermediate goods
$\tilde{n}_s$	Number of different firms in sector s
p	Price level
p_{m_i}	Price demanded by firm i in the manufacturing sector
p_j	Price of intermediate good j
p_s	Price of the good produced by type s labour
p_M	Price index of the manufacturing good
p_T	Price of the traditional good
$q_m(j)$	Quality of the $\bar{m}$th generation of the intermediate good j
$\tilde{q}_s(\tilde{j})$	Highest quality available of the component $\tilde{j}$
r	Interest rate
s	Skill index, $s \in \{L, H\}$ with L denoting low- and H high-skill
$\tilde{s}$	Worker index
t	Time index
$\bar{t}$	Time index
$\tilde{t}$	Time at which a job-worker pair is matched
u	Unemployment rate

u_s	Unemployment rate for labour of type s
v	Vacancy rate
v_s	Vacancy rate for labour of type s
w	Wage rate
w_{sS}	Wage rate paid to labour of type s in sector S
w^*	Equilibrium wage rate
$\hat{w}$	Wage rate measured in efficiency units
$\hat{w}^*$	Equilibrium wage rate measured in efficiency units
$\hat{w}^e$	Expected wage measured in efficiency units
$\hat{w}_i$	Wage rate measured in efficiency units paid by firm i
w^m	Monopoly union wage rate
w^{rtm}	Wage with right-to-manage wage bargaining
w_c	Wage rate in a perfectly competitive labour market
w_{m_i}	Wage paid by firm i in the manufacturing sector
w_i	Wage paid by firm i
w_s^*	Equilibrium wage rate for labour of type s
w_s	Wage rate for labour of type s
w_{s_j}	Wage rate in the intermediate sector for labour of type s
w_{sS}	Wage rate for workers of type s in sector S
w_S	Wage rate in sector S
$\bar{w}$	Alternative wage
$\bar{w}_s$	Alternative wage for labour of type s
$\hat{w}$	Wage rate measured in efficiency units
$\hat{w}_A$	Alternative income measured in efficiency units
$\breve{w}_s$	Proportional rate of change of the wage rate for type s
x_j	Quantity of intermediate good j
$x_s(i,\tilde{j})$	Demand of firm i in sector s for component $\tilde{j}$
y	Per-worker output
$\tilde{y}$	Marginal value of worker output
$\hat{y}$	Per-worker output measured in efficiency units

y_s Output per worker of type s

z Imputed real income of an unemployed

$\hat{z}$ Unemployment income measured in efficiency units

A Technology parameter

A_0 Initial productivity level

A_K Technology parameter for capital

A_L Technology parameter for labour

A_s Technology parameter for labour of type s

A_M Average quality of the intermediate goods

C Consumption level

D_s Production delay due to matching frictions

$\tilde{F}$ Fixed costs

$F(\bullet)$ Production function

$\hat{F}(\bullet)$ Production function in intensive form

G Household assets

I Investment

I_w Average wage income

K Capital stock

L Employment

$\bar{L}$ Total labour supply

$\bar{L}_S$ Total labour supply in sector S

L^* Equilibrium employment

L^D Labour demand

L^D_t Labour demand at time t

L^D_{LR} Long-run labour demand

L^{rtm} Employment with right-to-manage wage bargaining

L_0 Minimum employment level that unions will tolerate

L_{Lj} Firm low-skilled labour demand in the intermediate sector j

L_{m_i} Amount of labour employed by firm i in the manufacturing sector

L_s	Employment of labour of type s
$L_s(i)$	Amount of labour of type s employed by firm i
L_{sS}	Employment of labour of type s in sector S
L^S	Labour supply
L_S	Employment in sector S
L_U	Number of union members
M	Manufacturing good
P	Macroeconomic price index
Q_s	Average quality of the components in sector s
R	Revenue
R_j	Revenue of firm j
S	Sector index, $S \in \{M, T, X\}$, with M denoting the manufacturing, R the research sector, T the traditional sector and X the intermediate sector
T	Traditional good
U	Intertemporal utility
U_M	Unemployment in the manufacturing sector
V	Union utility
$\tilde{V}$	Intertemporal union utility
$\bar{V}$	Stock-market value of a monopolist in the intermediate sector
$\bar{V}_s(\tilde{j})$	Stock-market value of a monopolist in sector s producing component $\tilde{j}$
$\bar{V}^m$	Indifference curve with monopoly unions
$\bar{V}^{rtm}$	Indifference curve with right-to-manage wage bargaining
$\tilde{V}^E$	Lifetime utility of a currently employed worker
$\tilde{V}_s^E$	Lifetime utility of a currently employed worker of type s
$\tilde{V}^M$	Lifetime utility of a worker currently employed in the manufacturing sector
$\tilde{V}^T$	Lifetime utility of a worker employed in the traditional sector
$\tilde{V}^U$	Lifetime utility of a currently unemployed individual
$\tilde{V}_s^U$	Lifetime utility of a currently unemployed individual of type s

W^F Present-discounted value of expected profits from a filled position

W_s^F Present-discounted value of expected profits of a firm in sector s from a filled position

W^V Present-discounted value of expected profits from a vacant position

W_s^V Present-discounted value of expected profits from a vacancy in sector s

X Aggregate output of intermediate goods

$X_s(\tilde{j})$ Aggregate demand for component $\tilde{j}$ in sector s

Y Aggregate Output

Z^* Steady-state R&D input

$Z_s(\tilde{j})$ R&D input in sector s aimed at component $\tilde{j}$

α Labours' output elasticity

β Union-bargaining power

β_L Bargaining power of the union representing low-skilled workers

$\tilde{\beta}$ Worker-bargaining power

$\tilde{\beta}_s$ Worker of type s bargaining power

γ Elasticity of marginal utility

δ Rate of capital depreciation

$\epsilon_{f,\theta}$ Elasticity of the job-finding rate with respect to labour-market tightness

$\epsilon_{v,w}$ Elasticity of utility with respect to wages

ϵ_{v,w_L} Elasticity of utility with respect to low-skilled wages

$\epsilon_{F,L}$ Output elasticity with respect to labour

ϵ_{F,L_s} Output elasticity with respect to labour of type s

$\epsilon_{L,w}$ Elasticity of labour demand with respect to wages

$\epsilon_{v,\hat{w}}$ Elasticity of labour demand with respect to wages measured in efficiency units

ϵ_{L_Lj,w_L} Elasticity of low-skilled labour demand in firm j with respect to low-skilled wages

ϵ_{L_L,w_L} Elasticity of low-skilled labour demand with respect to low-skilled wages

ϵ_{L_m,w_m}	Elasticity of labour demand with respect to wages in the manufacturing sector
ε_1	Parameter of the effort function
ε_2	Parameter of the effort function
ε_3	Parameter of the effort function
ε_{s1}	Parameter of the effort function for type s labour
ε_{s2}	Parameter of the effort function for type s labour
ε_{s3}	Parameter of the effort function for type s labour
ε_4	Parameter of the effort function
$\hat{\varepsilon}_s$	Variable
ζ	Expenditure share on manufacturing goods
η_s	Sociological parameter
θ	Indicator of labour-market tightness
θ_s	Indicator of labour-market tightness for labour of type s
ϑ	Importance of wages as opposed to employment levels for unions
ι^*	Steady-state probability of a successful innovation
ι_j	Probability of a successful innovation in sector j
$\iota_s(\tilde{j})$	Probability of a successful innovation in sector s of component $\tilde{j}$
$\tilde{\iota}$	Total number of expected quality improvements
κ	Degree of homogeneity of goods and indicator for indicator of intensity of competition on the product market
λ	Innovation size
μ	Output coefficient in the research sector
ν	Worker utility
π	Firm profits
π^m	Profits with monopoly unions
$\bar{\pi}^m$	Isoprofits with monopoly unions
π^{rtm}	Profits with right-to-manage wage bargaining
$\bar{\pi}^{rtm}$	Isoprofits with right-to-manage wage bargaining
π_j	Profits of a market leader producing the intermediate good j

π_m	Profits in the manufacturing sector
π_{m_i}	Profits of firm i in the manufacturing sector
π_s	Profits of a firm producing an intermediate good of type s
π_0	Firms' fallback position
ρ	Rate of time preference
σ_{KL}	Elasticity of substitution between capital and labour
σ	Elasticity of substitution between two varieties of the composite manufacturing good
σ_s	Elasticity of substitution between low- and high-skilled labour
σ_S	Elasticity of substitution between low- and high-skilled labour in sector S
ς	Fraction of suitable applicants
τ	Time interval
ϕ	Fraction of workers who are skilled
φ	Production function in the research sector
$\chi_s(\tilde{j})$	Price of a component with quality $\tilde{q}_s(\tilde{j})$
$\psi(\bullet)$	Probability function
ω	Constant
Γ	Coefficient
$\hat{\Gamma}$	Coefficient
$\hat{\mathbf{\Gamma}}$	Matrix
Δ	Coefficient
$\hat{\Delta}$	Coefficient
$\mathbf{\Delta}$	Matrix
$\hat{\mathbf{\Delta}}$	Matrix
Θ_{sS}	Cost share of labour of type s in sector S
$\check{\Theta}_{sS}$	Proportional rate of change of the cost share
Λ	Lagrange multiplier
$\Pi_s(\tilde{j})$	Profits of a firm producing component $\tilde{j}$ in sector s
Υ	Number of vacancies within a firm

Φ	Coefficient
$\mathbf{\Phi}$	Matrix
$\Psi(\bullet)$	Distribution function
Ω	Nash product
$\mathcal{H}$	Hamilton function
$\mathcal{L}$	Lagrange function

1

Introduction

"There is a fact, a big unmistakable unsubtle fact: essentially every-where in the modern industrial capitalist world, the unemployment rates are much higher than they used to be two or three decades ago. Why is that? If macroeconomics is good for anything, it ought to be able to understand and explain this fact."
Solow (1986, p. S23)

Unemployment rates in all industrialised countries increased following the oil crisis in the seventies. In Germany, the "unification" shock also led to a substantial rise in unemployment. These increases following negative shocks are to be expected even in a well functioning labour market as workers and firms require time to adapt to the structural change. However, the disturbing feature is that these rates failed to decline in the long-term after the economy has had time to adjust to the new conditions. Therefore and unfortunately, the first part of the above quote is as true now as it was then: the unemployment rates in many industrialised countries are still much higher than they used to be in the 1960s and early 1970s which clearly indicates that the labour markets are characterised by persistent imperfections. Does that mean that macroeconomics is really not good for anything? Well, of course every reader is entitled to his or her own personal view on this question but fact is that the persistently high unemployment rates have led to a large increase in the amount of research on and understanding of the workings of labour markets. The following chapters do not try to summarise all of this research – perhaps the "Handbooks of Labor Economics" with their 5 volumes and more than 3600 pages come closest to achieving this – but aim to shed a little more light on three major branches of this research, the analysis of trade unions, efficiency wages and matching theories in order to gain deeper insights as to why countries are still struggling to reduce the number of people failing to find a job.

That unemployment is an economic, political and social problem which urgently needs to be tackled is without doubt. Simply by looking at the billions of euros that are transferred (and need to be financed) directly and indirectly from the government to the unemployed in the form of unemployment benefits, social insurance contributions, wage and tax subsidies, lost output, forgone tax income etc. makes clear how important it is to address this problem. However, when dealing with economic problems in general but perhaps particularly when it comes to talking about unemployment, it is especially important to bear in mind that we are dealing with a problem which is about much more than simply saving costs and increasing efficiency. In societies as we find them in all industrialised countries, people decide to work not only to earn income to be able to afford a standard of living above a national minimum level, but at least to a certain extent, enjoy the feeling of doing something productive, of being needed, of having influence etc., i.e. many individuals also want to work for numerous non-pecuniary reasons. Further, if a person becoming unemployed fails to "quickly" find a new job, there is also a danger that they have to give up their standard of living, circle of friends etc. and from a labour-market perspective, may lose some of the skills or human capital they previously possessed making it increasingly difficult for them to find employment in the future.

However, the fact that, at least in the short-term, individuals who become redundant retain their human capital is also one of the principle features by which the analysis of the labour market differs from the standard analysis of other markets. A standard car, for example, can be replaced by an identical one if need be and if competition is intense enough, the price of the car will equal its marginal production costs. A firm on the other hand can never "buy" or possess a worker; it can only rent the services the worker has to offer and use these to manufacture the actual good or service which the firm produces. Further, if an individual has certain characteristics that are important to a firm, then these characteristics cannot simply be transferred from one worker to the next but at best involve only a short training phase and at worst might even be unreplaceable. Such special skills may also give the worker some bargaining power vis-à-vis a potential or current employer, or if many workers with similar skills unite, make it possible for unions to exist and extract rents from employers. Naturally, by a similar argument it is often the case that firms also unite to employer confederations in order to try and secure as large a share of these rents as possible. It is for these reasons that one of the central themes of the following chapters will be the study of bargaining between unions and employers and on their consequences for the labour-market outcome. The existence of these rents which are bargained over also give first rudimentary insights into why labour market fails to clear in the sense that there is an unemployment rate of zero percent: it will only be possible to drive the wage down to market-clearing levels if firms are able to extract all the available rents.

Although the invention of computers, the internet, mobile phones etc. have made workers much more independent of a fix workplace, there will always be labour services which requires the worker to come to a certain location, for example, a factory. This requires that a worker lives relatively close to his workplace, a workplace which might not be his preferred living place and which may constrain the opportunities of other family members. This means that work conditions play a large role as they need to compensate for the loss in utility associated with living in an area which is not the ideal choice for the worker. Therefore, fairness interactions between employers and employees may well be important, a fact which was long neglected in both theory and practice. It is such considerations which provide the basis for the so-called efficiency-wage theories which provide the second major building block of the following chapters.

Finally, the labour market differs from other markets because, with workers and jobs needing to be relatively close together, labour will never be as mobile as goods which can easily be transported and sold around the world. The lower worker mobility also means that the labour market is characterised by more frictions than a standard goods or capital market, because workers (and firms) will only have incomplete information on all available opportunities in the market. Therefore, there is likely to be a delay in which it takes time to become informed about "suitable" opportunities, a fact which will also prevent the labour market from clearing and is analysed in more detail in the matching theories of unemployment, the third major building block of this volume.

As employers, employees, jobs, the general economic environment etc. each have a multitude of different characteristics, there are of course a huge number of aspects which potentially influence the performance of the labour market. Obviously, not all of these factors can be taken into account here, and, as is always the case in theoretical modelling, reality needs to be simplified to the extent that emphasis is only placed on those arguments which are of particular importance for the argument being made, and all other aspects need to be held as simple as possible or even ignored. Any other approach would simply lead to a specific model to become overloaded and run the danger that it becomes too complex to be analysed. Thus, there will be many features which are not analysed or only briefly mentioned.

For these reasons, taxation aspects will not be dealt with in any detail as both the theoretical and empirical evidence shows that they play a far more minor role than is often stated in public – see Pissarides (1998) or Nickell and Layard (1999). Similarly, although employment protection laws are often stated as being one of the prime causes of the poor continental European labour-market performance, the evidence that they have a decisive impact on unemployment rates is at best mixed – see Lazear (1990), Addison and Grosso (1996), Bentolila and Bertola (1990), Elmeskov et al. (1998) or Nickell and Layard (1999). Finally, even if especially ever since the 1990s there has been much public debate about the negative effects of "globalisation", studies on the impact of increased trade with emerging economies shows that the effects on wage and employment levels are small – see OECD (1997, Chap. 4).

Instead, as mentioned above, the following chapters focus on the theories of trade unions, efficiency wages and matching as these three approaches of labour-market theory best address the features by which the labour-market differs from other markets. As shown in more detail in Chapter 2 which provides a more detailed picture of the empirical evidence and so-called "stylised" facts in several OECD countries, there are three central questions which the theoretical explanations need to be able to answer: firstly, why do real wages not fall even in times of high unemployment? Secondly, what are the labour-market effects of the recent substantial rise in the supply of high-skilled labour and thirdly, how do conditions on the product markets influence the unemployment rate. These questions are analysed both in an environment with and without economic growth. Although economic theory by its own nature is not country specific, on a few occasions special reference will be made to the situation in Germany as this is the country where this volume was written.

Chapter 3 provides the basic theoretical building blocks for the following chapters. Hence, in this chapter, relatively rudimentary models of trade union behaviour, efficiency wages and matching theories are outlined. This means that only the labour market in one sector of the analysed economy is analysed in this chapter. This is done both for models in which labour is treated as homogeneous, and for the case in which labour differs with respect to skills.

In Chapter 4, the setup from the previous chapter is extended in two main respects: first, the labour market is now analysed in a general-equilibrium setting and second, an imperfectly competitive product market is introduced. This is done to highlight the fact that the labour and product markets are strongly interdependent, e.g. through a households' consumption and labour supply decision and from the firm perspective, labour demand is to a large extent derived from the demand for its products.

Chapters 5 and 6 both analyse the relationship between growth and unemployment in more detail. In Chapter 5 this is done under the assumption that labour is homogeneous, the source of growth is exogenous and perfectly competitive product markets. All these assumptions are relaxed in Chapter 6. Hence, there are two types of labour, low- and high skilled and the source of growth is through the endogenously determined creation of new higher quality products in a separate research sector. This final chapter is also concerned with the question as to why growth caused by improvements in technologies, has been more favoured towards the high-skilled.

Finally, Chapter 7 summarises the results and draws some final conclusions. Hopefully, here at the very latest, the reader has been convinced that macroeconomics is not only good for something but also essential in order to adequately analyse the reasons for the existence of long-term unemployment and to derive necessary policy consequences.

2

Empirical Evidence

That unemployment rates in the 1980s and 1990s have been persistently high especially in continental Europe is a well known fact. Whereas the following chapters will provide theoretical explanations for this development, this Chapter provides a more detailed description of the empirical observations. However, it needs to be noted throughout that there will always be several measurement methods for each question so that empirical evidence is seldom unambiguous. For example, although there is universal agreement in the fact that unemployment has risen, there is by no means agreement on how exactly unemployment should be measured and many different definitions of unemployment coexist. In the case of unemployment rates this is not only because of different scientific measurement concepts but also due to the fact that there is also a political interest at using concepts which "lower" the unemployment rate. Hence, even if it is clear that the unemployment rate is defined as the ratio of the number of unemployed to the labour force, there is no consensus as to which criteria need to be fulfilled by individuals in order for them to be counted as unemployed or as part of the labour force. For example, how intensive does an individual need to searching for a job to count as unemployed? In which age range and how healthy do they need to be? What counts as a job, i.e. how many hours a week does it involve? What does the unemployed need to have done before being officially registered at a job centre – e.g. students in Germany who are searching for their first job are not registered as unemployed. Similarly, there are just as many uncertainties as to who counts as part of the labour force? For example, some concepts count state employees whilst others do not, some contain the self-employed whereas in others this group is not included. That these different statistical concepts can lead to unemployment rates for one country and year which vary by several percentage points can be exemplified for the case of Germany. Thus, the German Federal Employment Services (Bundesanstalt für Arbeit) who neither include state and military employees nor the self-employed, calculate an unemployment rate for Germany for 2000 of 10,7% whereas the German Council of Economic

Experts who include the above obtain 9,6%. However, in order to be able to have internationally comparable rates, the OECD uses a unified concept for all countries to calculate their "standardised" unemployment rates. For Germany, the rate calculated according to this concept was (only) 7,9% in the year 2000. Figure 2.1 shows how the unemployment rates for several OECD countries measured according to this concept have developed since the mid 1960s. The figure clearly shows that all countries had "full" employment levels

Fig. 2.1: OECD Standardised Unemployment Rates

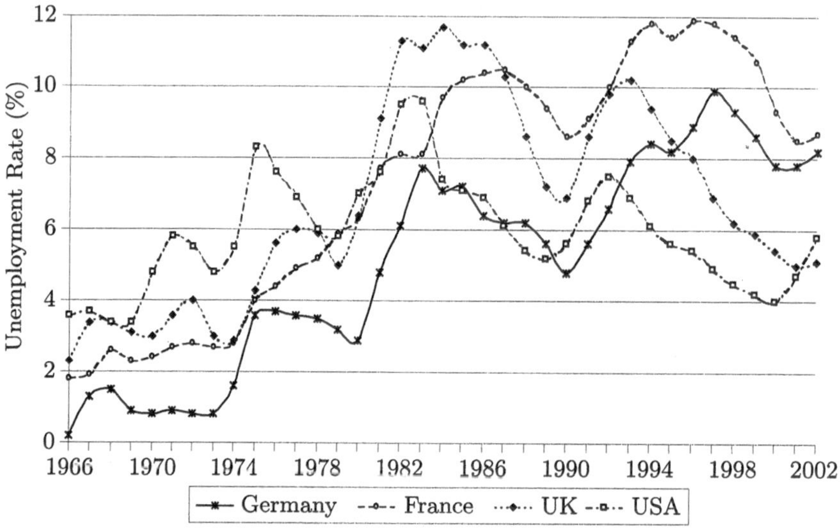

Source: OECD (1965–2001)

in the late 1960s and early 1970s, i.e. although there were positive unemployment rates these were so small that they could not be considered to represent a serious labour-market imperfection. However, following the oil price crises in 1973/74 and 1979/80, all countries experienced sharp increases in their unemployment rates. In Germany there was also a substantial jump in the rate after reunification in 1990. Although even in a well functioning labour market such increases in the rate of unemployment following negative shocks are to be expected, the figure also clearly shows that, at least in continental European countries, the unemployment rates failed to fall in the aftermath of these shocks, and increased again after each shock and it is this fact which gives cause for concern.

In a perfectly competitive labour market, an economist would expect the wage rate to fall as a consequence of the excess labour supply. However, as can be seen from Fig. 2.2 which plots the percentage change in real unit-labour costs for Germany, France, the United Kingdom and the United States, that with the exception of Germany in 1997 and 98 and the UK in 1994, wages never fell despite the high unemployment rates.[1]

Fig. 2.2: Change in the Unit-Labour Costs

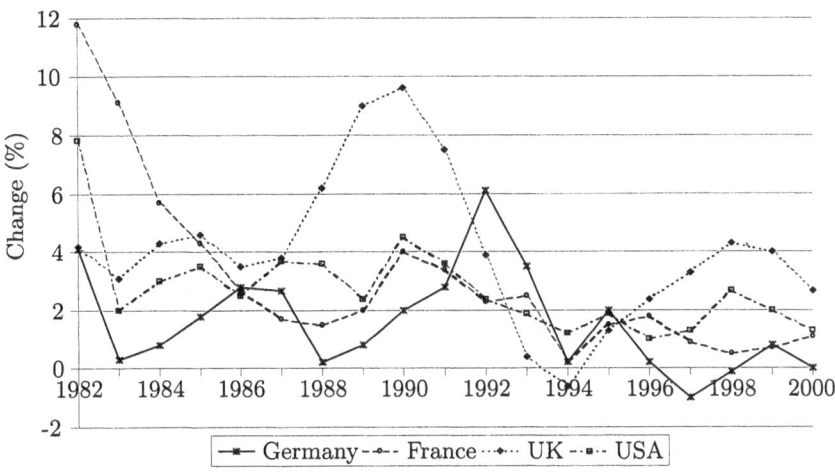

Source: OECD (1999, 2001a, Annex Table 13)

However, even if the unemployment and wage rates remain at relatively high levels, this does not imply that the labour market is "static" or completely inflexible. Firstly, all people switching between jobs without any unemployment spells will not show up in any unemployment statistics. Secondly, if the same number of people in any given time period become unemployed as there are unemployed finding new jobs, then the unemployment rate will remain constant no matter how large this number is. These dynamics can best be measured by looking at gross labour-market flows rather just net employment changes. Thus, as Fig. 2.3 shows for (West) Germany, a country whose labour market is always a prime culprit when it comes to talking about inflexibility, these flows by far outweigh the number of unemployed. Further, as can be seen from Table 2.1, the ratio of gross labour market flows to average stocks of (un)employed in Germany is similar to that in other countries including the United States and United Kingdom whose labour markets are always labelled as being the most flexible and highlights the large amount of job reallocation, i.e. simultaneous job destruction which is continuously taking place.[2]

---

[1] The average change in unit-wage costs between 1971 and 1981 were 5.7% in Germany, 11.7% in France, 14.3% in the UK and 7.5% in the USA and hence also all positive.

[2] See also OECD (1994a, Chap. 3) and Davis and Haltiwanger (1999) for overviews of international labour-market flows and of the different concepts as to how job flows should be measured.

Fig. 2.3: Gross Labour-Market Flows in Germany

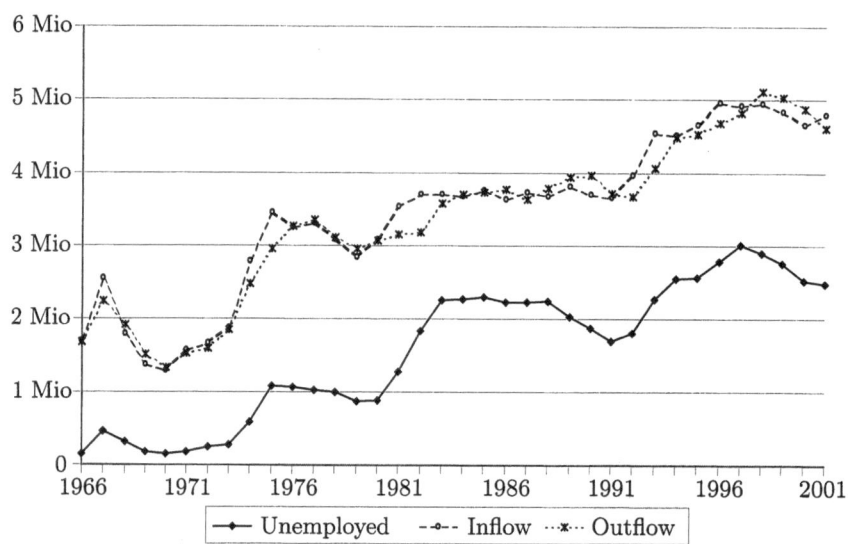

Source: Bundesanstalt für Arbeit (1965–2001)

Table 2.1: Gross Labour-Market Flows in Selected OECD Countries in 1987
(thousands)[a]

	Unemployment			Employment		
	Inflows	Outflows	Average Stock	Inflows	Outflows	Average Stock
France	4,115	4,128	2,728	4,528	4,841	15,685
Germany	3,726	3,636	2,479	6,046	5,811	27,070
U.K.	3,032	3,478	2,696	1,680	1,694	25,641
U.S.A.	19,770	20,227	8,312	27,077	28,432	107,150

[a] US data is based on labour-market surveys and therefore is not directly comparable with European data. US data refers to 1985. For France, employment flows (which cover establishments with more than 50 employees) include job to job reallocations.

Source: Burda and Wyplosz (1994, Table 1)

In any given time period, sector and country there will always be a certain degree of structural change caused, for example, by a change in consumer preferences, relative prices or by the development of new technologies and new products. Therefore, there will always be sectors in the economy or certain firms within a sector which are currently expanding and requiring more labour whilst others will be shedding labour. Hence, worker flows are closely related to job flows. This is shown in Table 2.2 which shows the figures for gross job gains (i.e. new firms opening and existing firms expanding) and gross job losses (i.e. firm closures and contractions) measured as average annual rates as a percentage of employment. The table also highlights the fact that the rate

Table 2.2: Comparison of Job Gains and Job Losses in Two Periods[a]

	Germany		France		UK[b,c]		USA[c]	
	1983-89	1989-90	1984-89	1989-92	1985-89	1989-91	1984-88	1989-91
Gross job gains	**8.7**	**10.2**	**13.9**	**13.7**	**9.1**	**8.0**	**13.2**	**12.6**
Openings	2.4	2.8	7.3	6.9	3.1	1.9	8.9	7.4
Expansions	6.4	7.4	6.6	6.8	6.0	6.1	4.3	5.1
Gross job losses	**7.7**	**6.6**	**12.8**	**13.9**	**6.7**	**6.4**	**10.0**	**11.1**
Closures	2.0	1.8	6.9	7.1	4.2	3.4	7.2	7.6
Contractions	5.7	4.8	5.9	6.8	2.5	3.0	2.9	3.5
Net Employment Change	**1.1**	**3.6**	**1.2**	**-0.2**	**2.4**	**1.6**	**3.2**	**1.4**
Net entry (openings *less* closures)	0.4	1.0	0.5	-0.2	-1.1	-1.5	1.7	-0.1
Net expansion (expansions *less* contractions)	0.6	2.6	0.7	-0.1	3.5	3.1	1.4	1.6
Job turnover	**16.4**	**16.8**	**26.7**	**27.6**	**15.8**	**14.4**	**23.2**	**23.7**

[a] Sampling months/periods vary across countries as follows: Germany: June; France; United Kingdom: December; United States: December (June in 1989 and 1991).

[b] Data refers to firms

[c] See notes as to the quality of the data in OECD (1994a, p. 108)

Source: OECD (1994a, Table 3.4)

of structural change measured by the job-turnover rate (i.e. the sum of gross job gains and losses) is of similar magnitude in both France and Germany on the one hand, whose labour markets are often said to be "rigid", and the United States and United Kingdom on the other hand, the two countries whose labour markets are often seen as exemplary.[3] The second feature which stands out from Table 2.2 is the fact that the job-turnover rate is increasing in all countries with the exception of the United Kingdom, i.e. the pace of structural change is becoming faster.

The fact of relatively high job-turnover rates and the simultaneous increase in the unemployment rate can be best understood by noting that there are two components which determine the unemployment rate: the probability or risk of becoming unemployed and the duration of unemployment. Thus, at the two extremes, an unemployment rate of 5% can be caused either by all individuals in an economy being unemployed for 5% of the year, or by 5% of all individuals being unemployed the whole year. The relationship between the risk (measured as the ratio of inflow into unemployment to number of employees) and duration (in weeks) is calculated by the German Federal Employment Services and is shown in Fig. 2.4. This figure clearly shows that in

Fig. 2.4: Unemployment Risk versus Duration

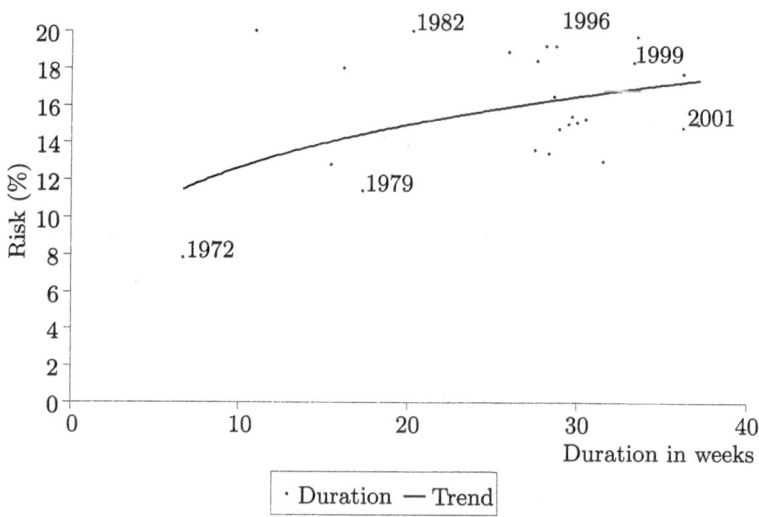

Source: Bundesanstalt für Arbeit (2003, Overview 3.2.1)

Germany it is above all the higher unemployment duration which has contributed to the increase in unemployment whereas the (average) risk of becoming

---

[3] In fact, Sweden and Italy both have higher job turnover rates. However, New Zealand which undertook many far-reaching labour-market reforms aimed at creating a more flexible labour market in the 1990s had a much larger turnover rate of 37.4 between 1987-89 although this rate has subsequently decreased slightly.

unemployed has remained more or less constant. The same also holds for France as can be seen from Table 2.3. However, as can also be seen from the table, this increase in unemployment duration is one of the empirical stylised facts in which the continental European labour markets differ from those in the United Kingdom and United States where long-term unemployment (i.e. an unemployment spell lasting more than one year) and also unemployment durations of between six months and one year have actually decreased.[4]

A further indicator of the pace of structural change in an economy can be obtained by looking at the number of vacancies and unemployed at any moment in time. That both exist simultaneously was first investigated by Beveridge (1944). He found a convex to the origin relationship between the two – a fact which now bears his name and is depicted as a Beveridge curve. This indicates that the marginal returns to firms of opening up additional vacancies, i.e. the rate at which firms can find "suitable" employees, decreases with the number of already existing vacancies. The intuition behind this result is that if there are already many vacancies, then the currently unemployed will find it easier to obtain jobs in any given time period, lowering the number of unemployed. This means that there are less people searching for a job so that it now becomes more difficult and time consuming to find an employee. Similarly, the lower the number of vacancies, the more difficult it will be for the unemployed to find jobs, and hence the higher is the unemployment rate. However, the fact that there are both vacancies and unemployed at any given time implies that firms are not willing to accept all potential employees. Thus, if due to structural change the job requirements, for example with respect to skill, do not "match" the skills a potential employee has to offer, then both the firm and the unemployed will continue searching. Hence, the number of vacancies and unemployed at any moment in time are an indicator of the degree of "mismatch" in the labour market. The fact that unemployment duration has increased in some countries is an indicator that this mismatch has increased over time in which case the Beveridge curve shifts outwards. That this has in fact occurred can be seen from Fig. 2.5 which shows that the Beveridge curves for Germany and France have clearly shifted outwards, whilst the curves for the United Kingdom and United States have shifted inwards recently. However, the figure also clearly shows that the negative relationship between unemployment and vacancies discussed above does not hold at all times.

The increasing job-turnover rate shown above is an indicator that there are many new jobs being created in all economies at any point in time. However, the fact that both the (average) unemployment duration has increased and in some countries the Beveridge curve has shifted outwards leads to the conclusion that there must be a certain group of individuals who are now much more affected by unemployment than they were in the past and exemplifies how unemployment can polarise the society into two groups: one with only

---

[4] See OECD (2001c) for details on other OECD countries.

Table 2.3: Unemployment Duration [a]

	1983	1984	1985	1986	1987	1988	1989	1990	1991	1992	1993	1994	1995	1996	1997	1998	1999	2000
Less than one month																		
Germany	6.1	6.8	6.3	5.3	5.3	5.8	6.0	9.2	10.6	12.3	10.7	7.6	7.3	6.5	5.8	6.8	6.0	
France	5.2	4.9	4.4	5.2	4.9	5.4	6.1	4.4	4.8	4.2	4.1	3.7	4.1	3.7	3.8	7.0	3.8	
UK	6.8	7.2	6.7	6.8	6.9	9.4	11.2	12.5	11.5	9.7	10.2	10.6	11.2	13.6	14.2	13.5	15.5	
USA	33.3	39.2	42.1	41.9	43.7	46.0	48.6	46.3	40.4	35.1	36.5	36.4	37.7	42.2	43.7	45.0		
More than one month and less than three months																		
Germany	11.2	12.3	11.9	10.8	11.8	12.0	12.4	12.3	14.8	11.0	9.0	8.0	10.6	9.4	9.2	10.9	11.7	
France	11.2	11.9	9.5	10.0	12.5	12.9	12.8	18.8	17.3	16.3	14.1	14.0	14.7	14.5	13.9	17.9	16.7	
UK	10.7	11.9	10.1	11.1	11.4	14.2	15.7	17.5	19.3	13.2	11.2	12.4	12.7	14.5	16.2	20.1	22.4	22.4
USA	27.4	28.7	30.2	31.1	29.6	29.9	30.3	32.0	32.4	28.9	30.1	31.6	31.6	31.7	31.4	31.2	31.9	
More than three months and less than six months																		
Germany	16.8	16.0	15.3	16.6	19.1	17.9	15.2	17.1	21.8	23.0	18.6	17.6	15.9	17.9	15.8	15.2	15.1	14.8
France	16.7	16.7	15.1	14.7	16.2	17.1	17.0	19.7	19.0	19.8	21.3	20.1	18.3	19.8	18.0	18.1	19.6	17.6
UK	16.1	15.7	15.8	16.7	16.9	16.5	17.7	19.8	22.1	18.2	16.1	14.0	16.0	16.2	15.4	18.4	18.7	18.8
USA	15.4	12.9	12.3	12.7	12.7	12.0	11.2	11.7	14.4	15.1	14.5	15.5	14.6	14.5	14.7	12.3	12.8	11.8
More than six months and less than one year																		
Germany	24.2	20.4	18.8	19.1	15.5	18.1	17.3	17.9	22.6	21.9	19.7	19.5	17.2	17.5	18.3	17.0	15.4	16.1
France	24.7	24.3	24.2	22.2	20.9	19.8	19.9	17.5	20.8	22.0	24.0	23.3	21.7	22.0	22.5	20.1	15.2	19.5
UK	20.8	18.9	17.1	17.2	16.9	16.8	16.2	15.9	18.3	21.9	20.4	18.0	17.2	18.3	16.1	14.6	15.8	15.2
USA	15.4	12.9	12.3	12.7	12.7	12.0	11.2	11.7	14.4	15.1	15.5	14.6	14.5	14.7	12.3	12.8	11.8	
More than one year																		
Germany	41.6	44.5	47.8	48.3	48.2	46.2	49.1	46.8	31.6	33.5	40.3	44.3	48.7	47.8	50.1	52.6	51.7	51.5
France	42.2	42.3	46.8	47.8	47.8	45.5	44.8	43.9	38.0	37.2	36.1	38.3	42.3	39.5	41.2	44.1	40.3	42.5
UK	45.6	46.3	50.3	48.2	47.9	43.0	39.1	34.4	28.8	35.4	42.5	45.4	43.6	39.8	38.6	32.7	29.6	28.0
USA	13.3	12.3	9.5	8.7	8.1	7.4	5.7	5.5	6.3	11.1	11.5	12.2	9.7	9.5	8.7	8.0	6.8	6.0

[a] As a percentage of unemployment

Source: OECD (2001c)

Fig. 2.5: Beveridge Curves

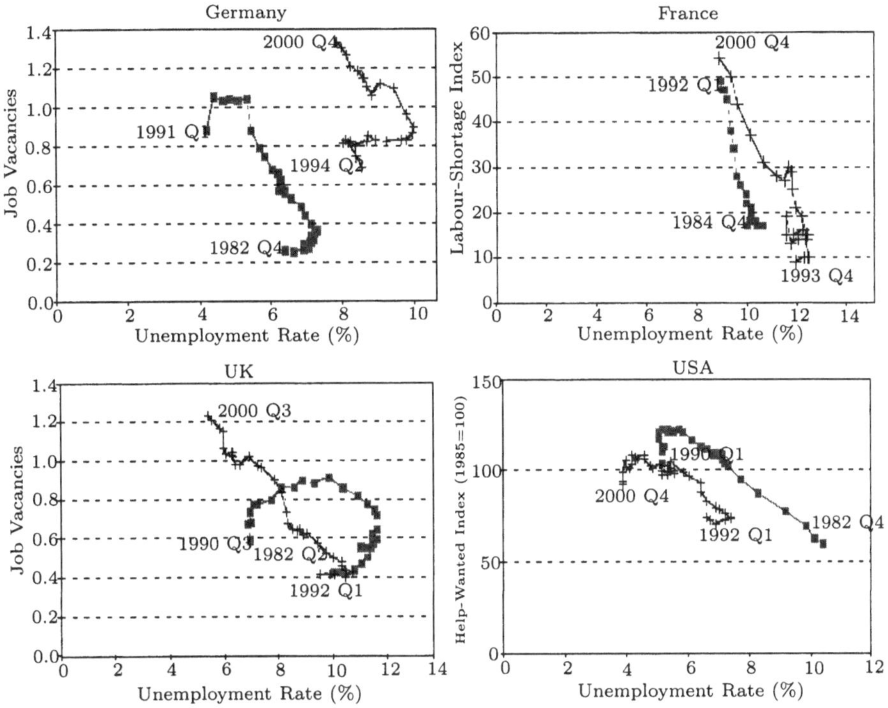

Source: OECD (2001b, p. 18pp.)

short periods of unemployment and the other who, should they become unemployed, need a long time to find a new job. As the unemployment rate is always an average over all sectors and types of individuals, it can easily be the case that one group of individuals experiences an increase in the incidence of unemployment whereas another group has less difficulties in finding a job, leaving the macroeconomic unemployment rate unaltered. That this is indeed the case can be seen by looking at the unemployment rates disaggregated with respect to the educational qualifications of the workforce as shown in Table 2.4.[5] As can be seen from the table, with the exception of France in the mid to late 1990s, the unemployment rates of the low-skilled are at least twice as high as those of their high-skilled counterparts. In fact, in Germany they were almost five times as high in the early 1990s. Further, the ratio of low- to high-skilled unemployment has been increasing in all the listed countries since 1994 with the exception of the United Kingdom in 1996 and the United States in 1995 and 1998. Finally, although the ratios are all similar, the absolute levels of the low-skilled unemployment rates are much lower in the UK and the US relative to their values in Germany and France (the ratio between France and the US is 1.8 in the year 2000).

---

[5] For a more detailed analysis for Germany see Reinberg (1999).

Table 2.4: Unemployment Rates by Educational Attainment

	1971–82	1983–90	1991–93	1994	1995	1996	1998	1999	2000
Germany									
Low-Skilled	6.4	13.0	10.7	13.9	13.3	14.2	16.6	15.8	13.7
High-Skilled	1.7	3.1	2.2	7.0	6.3	6.9	8.0	6.7	5.8
Ratio	3.8	4.2	4.9	2.0	2.1	2.1	2.1	2.4	2.4
France									
Low-Skilled	6.5^a	9.9	12.1	14.7	14.0	14.8	14.9	15.3	13.9
High-Skilled	2.1^a	2.6	4.2	8.6	7.7	8.2	8.0	7.7	6.5
Ratio	3.1	3.8	2.9	1.7	1.8	1.8	1.9	2.0	2.2
UK									
Low-Skilled	7.5	15.9	17.1	13.0	12.2	10.9	10.5	10.0	8.9
High-Skilled	2.4	4.4	6.2	6.0	5.5	5.2	3.8	3.7	3.3
Ratio	3.1	3.6	2.6	2.2	2.2	2.1	2.8	2.7	2.7
USA									
Low-Skilled	7.8	11.3	11.0	12.6	10.0	10.9	8.5	7.7	7.9
High-Skilled	2.0	2.4	3.0	4.6	3.8	3.7	3.2	2.9	2.7
Ratio	3.9	4.7	3.7	2.7	2.6	3.0	2.7	2.7	3.0

[a] Data refers to 1982 only

Sources: Nickell and Bell, Table 2a, 1996, Table 1 and own calculations
        based on OECD (1997–2002, Tables D).

However, lower unemployment rates alone do not imply that the low-skilled
fare better in the UK and US. As can be seen from Fig. 2.6 which plots the

Fig. 2.6: Trends in Earnings Dispersion
Log of Ratio of Wage of 90th to 10th Percentile Earner

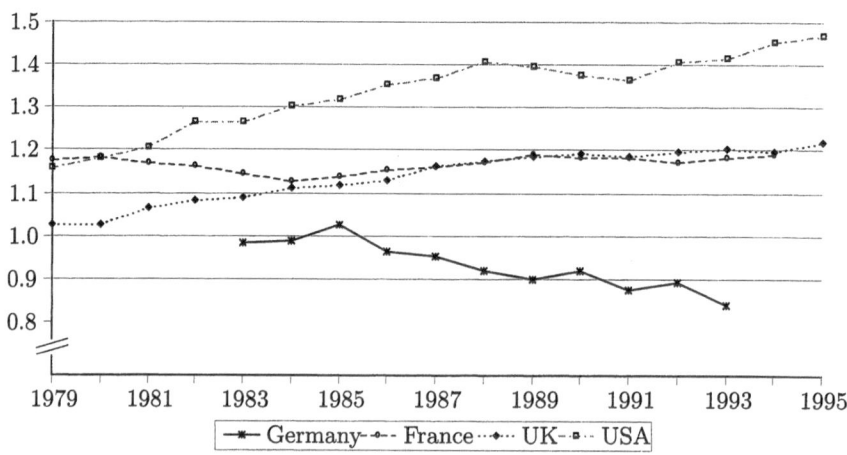

Note: Data for USA refers to males only

Source: OECD (1996, Table 3.1)

upper earnings limits of the ninth relative to the first decile, the relative wages
of the low-skilled have fallen much more in absolute and relative terms in the

United States and United Kingdom relative to the wage differential in France and Germany.[6] Whereas in the USA the log of earnings of the ninth relative to the first decile increased by 0.31 between 1979 and 1995, it only increased by 0.01 in France in the same time interval and actually decreased in Germany (between 1983 and 1994). That the wage differential has increased in all the above countries with the exception of Germany is all the more surprising given the fact that, starting in the 1970s, the relative supply of high-skilled workers has increased substantially as shown in Table 2.5. Further, despite

Table 2.5: Relative Supply of Population with a Tertiary-Education Degree[a]

	1969	1979	1989	1992	1994	1995	1996	1998	1999
Germany	6.0^b	7.4^c	9.4	12.0	13.0	13.0	13.0	14.0	13.0
France	5.3^d	8.3^e	11.8	10.0	9.0	11.0	10.0	10.5	11.0
UK	8.0^f	12.0	18.3	11.0	12.0	12.0	13.0	15.4	17.0
USA	10.8^g	16.6^g	21.5^g	24.0	24.0	25.0	26.0	26.6	27.0

[a] Age 25–64   [b] 1976   [c] 1982   [d] 1970   [e] 1980
[f] 1973   [g] Age 18–64

Sources: OECD, Table 5.7, 1995-2001, Tables A2.1a

this increase in the share of high-skilled workers, they have also been able to increase their employment shares, i.e. the demand for these workers has increased even more than the supply as shown in Table 2.6. This indicates

Table 2.6: High Education Employment Shares

	1977	1981	1985	1989
Germany	.032	.044	.054	.066
France	.047	.057	.081	.093
UK	.039	.054	.065	.064
USA	.088	.126	.161	.167

Source: Machin and van Reenen (1998, Table I)

that there has been "skilled-biased technological change" which will be looked at in more detail in Chapter 6.

The fact that the unemployment rates and (relative) wage levels are very different depending on the educational status highlights the fact that there are indirect aspects affecting labour-market performance, e.g. the research system of a country. However, one indirect effect that is often overlooked in the literature and public debate on policies designed to reduce the level of unemployment is the role that product-market imperfections play in determining the labour-market outcome.

As shown in Fig. 2.7 which plots industry wage premia against industry-specific indicators of product-market regulation, there is a strong and significant

---

[6] For a more detailed analysis for Germany see Fitzenberger (1999).

Fig. 2.7: Wage Premia and Product-Market Regulation[a]

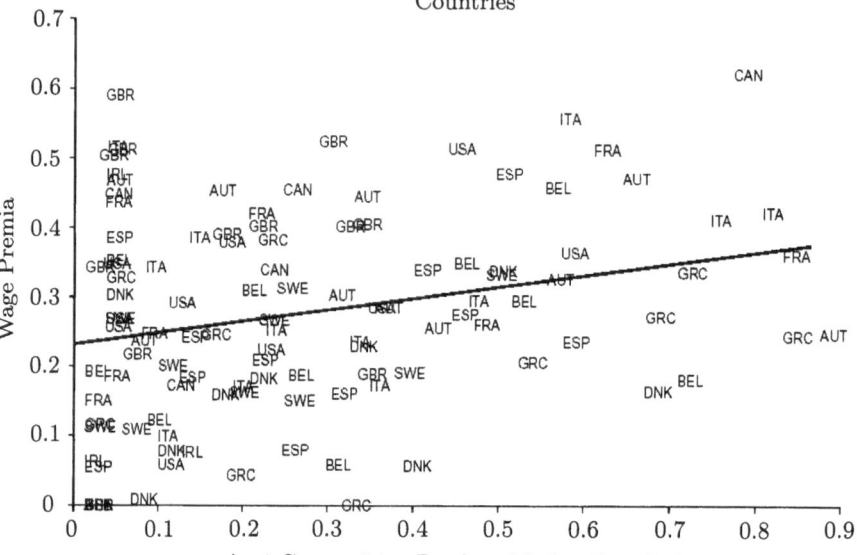

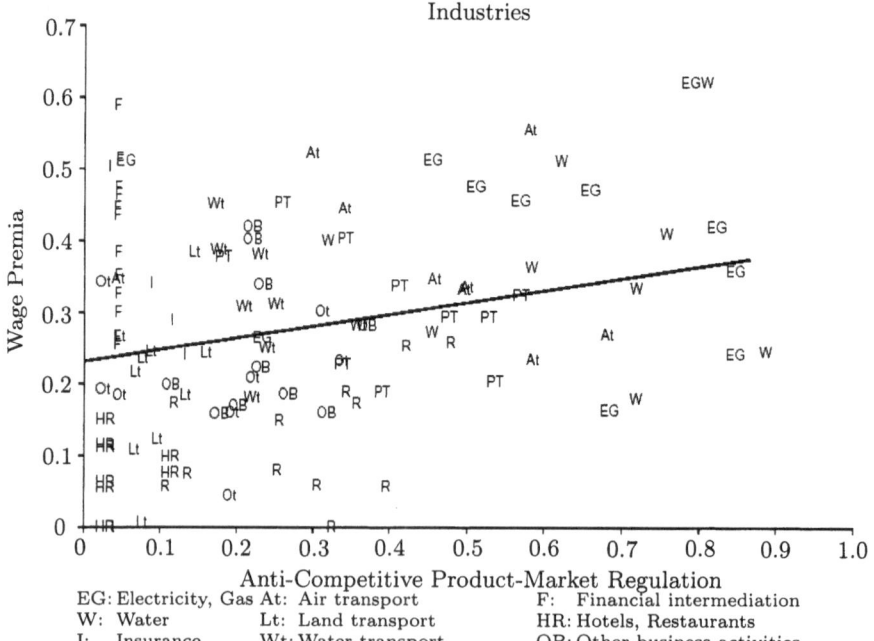

a Data refers to 1998

Source: Jean and Nicoletti (2002, p. 24)

correlation between higher product-market regulation, i.e. lower product-market competition, and the wage premia (defined as the logarithm of the industry wage relative to the economywide, employment weighted, average wage) across all countries and industries. The notion that increased competition on the product market leads to lower rents and thus also to lower wage premia (and hence lower unemployment) was first investigated by the OECD in its 1994 Jobs Study (see OECD 1994b, p. 23pp.). Since then, many empirical tests, summarised in OECD (2002, Chap. 5), have found convincing evidence in favour of this hypothesis. There are numerous channels through which product-market conditions are transmitted to the labour market. As will be shown more formally in Chapter 4, higher product-market competition raises the elasticity of product demand which in turn raises output and hence labour demand. Further, if more intense competition created through product-market deregulation has the side effect that market barriers to entry are reduced or even eliminated, then potential new firms will find it easier to establish themselves which again leads to higher demand elasticity (see Blanchard and Giavazzi 2001 for evidence). Such a fall in the barriers to entry in a certain industry or sector may also lead to an increased supply of capital to this branch of the economy (see OECD 1998 and Fonseca et al. 2002 for evidence). Finally, to the extent that higher product-market competition tends to dissipate rents, there is naturally less scope for rent-seeking behaviour by both employers and employees which has the effect of reducing wage premia in a certain sector. Not only does such a reduction in the wage premia lead to higher labour demand, but such a decline in the wage relative to that paid in other sectors will also lead some workers to give up queuing or "waiting" for such a high-wage job (see Kletzer 1992 for evidence).

As lower product-market competition leads to higher wage premia, it also has consequences for the employment level. Figure 2.8 plots the employment rates (defined as the share of the working-age population which is employed) as a function of a summary product-market competition index (see Nicoletti et al. 2002 for a detailed description of how this index is determined).

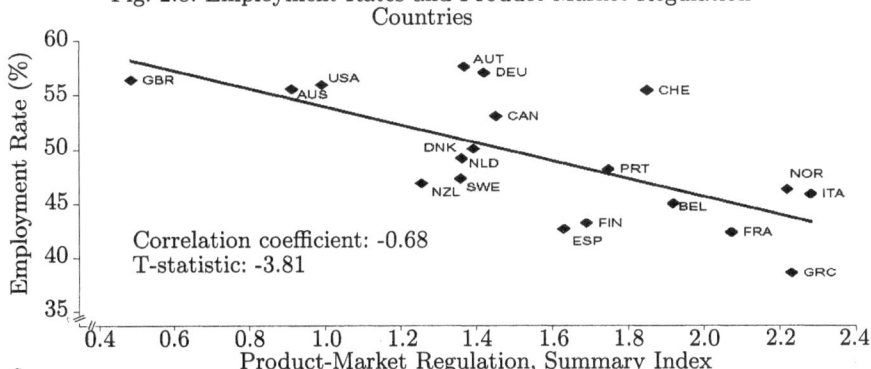

Fig. 2.8: Employment Rates and Product-Market Regulation[a] Countries

[a] Data applies to 1998 for the non-agricultural business sector
Source: Nicoletti et al. (2002, p. 47)

As can clearly be seen from this figure, a more regulated product market which lowers the intensity of competition also leads to lower employment rates. The figure also shows that the average degree of product-market regulation is lower in the United Kingdom (which, according to this index, is the most "liberalised" country) and the United States than in Germany and France.

Apart from the influence of product-market competition intensity on the labour market, the relationship between growth and unemployment is also analysed in more detail in the following chapters. Empirically, as shown in Fig. 2.9 which plots the growth rates of real GDP against the standardised unemployment rates, there are no clear cut results on the link between these two variables.

Fig. 2.9: The Relationship Between Growth and Unemployment Rates

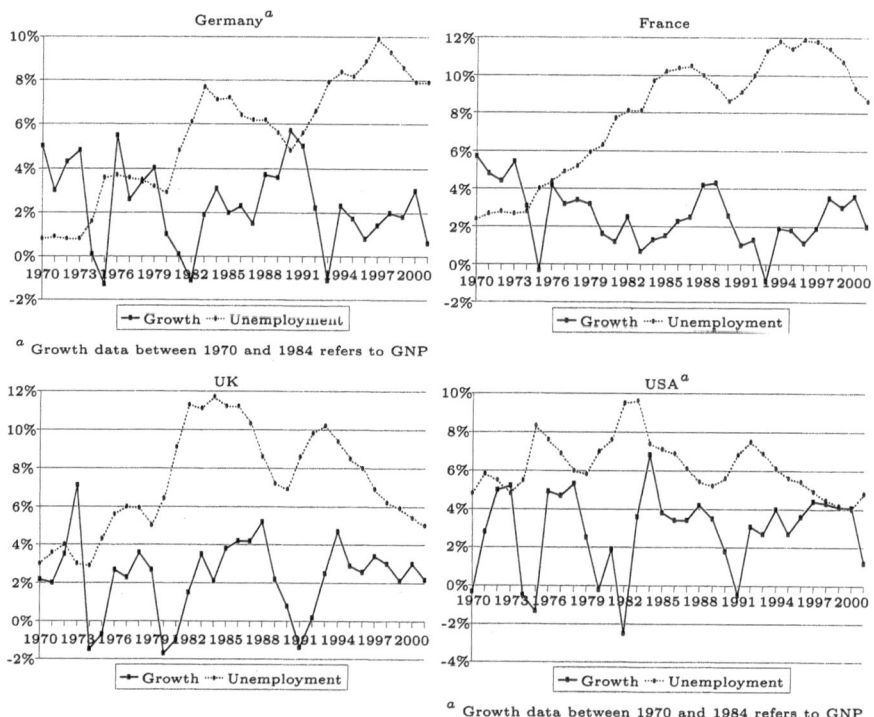

Source: OECD (1987–2002, Tables R.1 and Annex Table 1)

For example, in Germany the change in growth and unemployment rates was positively correlated in 1976, 77, 80, 83, 87, 89, 94, 95, 97 and in the year 2000, but negatively correlated in the other time periods. Similar results also hold for France, the United Kingdom and the United States.

That the relationship between growth and unemployment is similar in the four OECD countries listed above highlights the fact that there are both common developments in continental Europe and the UK and USA, but that the two "regimes" also differ in some aspects. Thus, whereas the unemployment rates have remained stubbornly high in France and Germany relative to the levels in the United Kingdom and United States, the gross labour-market flows are similar. Further, whereas the unit-wage costs have increased in Germany almost continuously since 1965, and the differential between the low- and high-skilled wage has remained nearly constant in France and decreased in Germany, it has increased substantially in the Anglo-American countries in spite of a large increase in the supply of high-skilled labour. Further, in all of the above-mentioned countries there is a negative relationship between the unemployment and vacancy rates in most time intervals. However, this so-called Beveridge curve has shifted outwards over time in Germany and France but has tended to shift inwards on the other side of the Atlantic and Channel.

Thus, the focus of the models in the following chapters will be to explain why wages do not fall even in times of persistent high unemployment rates, how the existence of Beveridge curves can be explained and on the differences in the development of the wage differential. These labour-market models will be analysed in detail in the following chapters.

3

Wage Setting and Sector-Specific Unemployment

One of the main facts highlighted in the previous chapter is that, with very few exceptions, neither the nominal nor the real wage level has fallen even in times of high unemployment. Therefore, any theory of unemployment must be able to explain which mechanisms prevent wages from falling to market-clearing levels. The three most predominant theories which explain such wage behaviour found in the literature are models of union wage bargaining, efficiency-wage theories or matching models and it is for this reason, that in line with this mainstream literature, the focus here will be on these three classes of models.[1] The purpose of this chapter is to provide introductory analysis of these theories. Although this has the disadvantage that many realistic aspects are ignored and many simplifying assumptions need to be made, it has the advantage that the main concepts behind these models can be better illustrated and that a unified framework is developed which can then subsequently be modified. In this rudimentary setting, the models developed in this chapter will first analyse the effects of unions, efficiency wages and matching friction for the case that labour is treated as homogeneous, and then subsequently for the case that labour is heterogeneous with respect to skills. These frameworks provide the foundations for more complex and realistic settings developed in the subsequent chapters.

Section 3.1 analyses the effects on wages and employment when wages are set through bilateral bargaining between unions and employers. It will be shown that, as soon as economic rents exist, that the union, acting as a monopolist

---

[1] Less common is the theory of implicit contracts which analyses the effects of long-term relationships between employers and employees. This theory dates back to Baily (1974), Gordon (1974) and Azariadis (1975) – see Rosen (1985) for a survey. However, as implicit contracts are rarely found in practice and these models make counterfactual predictions, e.g. that employment should be increased in recessions, this class of models will not be discussed here. The slightly more common "insider-outsider" models will be discussed in the context of union wage-bargaining models in the next section.

with respect to the supply of labour, will be able to force the employers to share some of these rents with their workforce by accepting lower profits and paying wages above market-clearing levels. As unions act on behalf of employed members in specific sectors of the economy, their wage demands need not necessarily fall even if unemployment increases.

Section 3.2 discusses the class of models known as efficiency-wage theories. Although there are many different types of these theories, they have in common that they all explain possible incentives firms have of paying their workers higher wages. Such a higher wage can, for example, reduce labour turnover and therefore training costs, save the firm monitoring costs or have productivity enhancing effects if workers exert more effort (and thus more output) with higher wages. Hence, paying a higher wage implies both a cost and a benefit to the firm and the optimal wage, i.e. the wage that minimises unit labour costs, need not necessarily be the market-clearing wage.

Finally, in Section 3.3, labour-market imperfections in the form of search and matching frictions are introduced. The central concept behind this class of models is that neither firms nor workers are perfectly informed about job-seekers or job-offers respectively, but must first undergo a costly search process before a job-worker pair can be brought together or "matched". These frictions on the labour market mean that rents accrue as soon as a job-worker pair is found as both sides of the labour market can benefit by forming a match and thus saving each other future search costs. For this reason, similar to models of union wage bargaining, the worker has a certain degree of bargaining power and can therefore prevent the wage rate from falling to market-clearing levels.

3.1 Union Wage Bargaining

Historically, unions evolved to form a counter-power to prevent firms from solely dictating working conditions and being able to exploit their workforce. Although nowadays numerous government regulations exist in industrialised democracies which help guarantee minimum work standards, the notion that if all workers join together to form a union they have more bargaining strength and can better promote their interests than an individual worker could on his own, still holds today. Hence, unions aim to improve the material welfare of their members in general, primarily by trying to bargain for higher wage levels. Even if the importance of unions undoubtedly varies across countries, they still play a predominant role in most of Western Europe, with the notable exception of the United Kingdom where their bargaining strength was drastically reduced during the Thatcher years (1979–1990) and has remained low ever since.

There are of course numerous measures for the "importance" of unions. Looking solely at membership figures, then with the exception of Finland and

Sweden, ever since the seventies, unions have been losing members in all of Europe. This can be seen from Table 3.1 which shows how union density, defined as trade union members as a percentage of all wage and salary earners, has evolved since 1970 for a number of European countries as well as the United States. Judging by this measure alone, apart from the Scand-

Table 3.1: Union Density and Coverage in Selected OECD Countries

Country	Density (%)				Coverage (%)		
	1970	1980	1990	1994	1980	1990	1994
Austria	62	56	46	42	98^a	98	98
Belgium	46	56	51	54	90^a	90	90
Denmark	60	76	71	76^b	69^a	69	69
Finland	51	70	72	81^c	95	95	95^c
France	22	18	10	9	85	92	95^c
Germany	33	36	33	29^b	91	90	92
Italy	36	49	39	39^d	85	83	82^b
Netherlands	38	35	26	26^b	76	71	81
Norway	51^e	57	56	58	75^a	75	74^b
Portugal	61^f	61^g	32	32^h	70	79	71^b
Spain	27^i	9	13	19	76^a	76	78
Sweden	68	80	83	91^b	86^a	86	89
Switzerland	30^j	31	27	27^d	53	53	50
UK	45	50	39	34	70	47	47
USA	23^i	22	16	16	26	18	18

^{a} Data for 1980 was not available. 1990 data used instead
^{b} 1993 ^{c} 1995 ^{d} 1992 ^{e} 1972 ^{f} 1978 ^{g} 1984 ^{h} 1990
^{i} 1977 ^{j} 1971

Sources: OECD (1994a, Chap. 5), OECD (1997, Chap. 3)

inavian countries, unions do not seem to play an important role with density values in 1994 ranging from 9% in France to 54% in Belgium. However, a far more accurate indicator of union influence is the number of workers whose wage contract is covered by a collective contract between employers and unions. This influence is measured by the "coverage" rate, which is also shown in Table 3.1.[2] As can be seen from the table, this value is much higher in all countries. Again, looking at continental Western Europe, the lowest union coverage is in Switzerland where 50% of the workers are covered, and the highest coverage is in Austria where 98% of all workers are covered. The difference between union density and coverage can most clearly be seen in France, which

---

[2] See OECD (1994a, Chap. 5) for coverage rates which differentiate amongst others, by industry, public versus private sectors and firm size.

has the lowest density of all countries listed (including the United States), but has the joint second-highest coverage. Within Europe, the United Kingdom has the lowest coverage rate which is still almost three times as high as the corresponding value for the United States. With these obvious institutional differences between the United States and continental Western Europe, union influence is one of the most frequently named reasons for the high (and persistent) unemployment rates observed in Europe.

As discussed above, one of the ways a union can act in the interests of its members is by pushing through higher wages. This higher wage rate provides workers with an inherent interest in becoming a union member.[3] Although we ignore how precisely objectives are formed *within* a union, we assume that the union is characterised by precisely defined preferences and acts rationally on behalf of its members.[4] Therefore, assuming that individual worker utility is a positive function of the wage rate, then the union utility (or objective) function itself also positively depends on the wage level it negotiates for its members. In reality, utility and thus wage negotiations will not only bargain over the wage level but also cover a multitude of aspects, including amongst others, work organisation, quality of output and more recently worker on-the-job training.[5] These aspects are ignored here and in line with the vast majority of the literature, emphasis is placed on wages and employment so that only these two factors are assumed to influence union utility. However, even if it is clear which arguments are to enter the utility function, how they enter the union objective function depends on the underlying utility concept as to how the utility of the individual union members influences union welfare.

One possible concept is the so-called "utilitarian" welfare function. If any disutility workers might have from providing effort is ignored for the moment, leaving this discussion until Section 3.2, then in this case union utility V is simply the sum of the utility of the individual members, i.e.

$$V = \sum_{\tilde{s}=1}^{L} \nu_{\tilde{s}}(w) \tag{3.1}$$

where w denotes the wage rate of the $\tilde{s}$ workers, L is employment and $\nu_{\tilde{s}}(w)$ is the utility that worker $\tilde{s}$ derives from wages with $\nu'_{\tilde{s}}(w) > 0$.

---

[3] However, higher wages need not be the sole motive for a worker to join a union. See Agell (1999, 2002) and Agell and Lommerud (1992) who analyse possible insurance incentives for the worker.

[4] Although this assumption is by now completely standard in the theoretical literature on unions, there was a famous debate in the forties between Dunlop (1944) who argued that unions do have objective functions and Ross (1944) who placed more emphasis on the internal structure of unions.

[5] See, for example, Booth (1995) or Sapsford and Tzannatos (1993) for a more detailed discussion of different aspects of union aims and the effects of unions on variables other than wages and employment.

As can be seen from (3.1), the utility of each individual worker is independent of the utility of other workers. This is without doubt a critical assumption as it means that the union simply represents many individual preferences. In reality, it is far more likely that the union will act more like a political party and position itself according to its "median voter", i.e. act so as to maximise the number of its members. However, seeing as we are mainly concerned with the effects of union bargaining on wage levels, this assumption will not prove to be critical.[6] Further, an underlying assumption in the utility function as given by (3.1) is that all workers receive the same wage. This assumption is not problematic if workers are homogeneous (as in Chapters 3.1.1, 4.2.1 and 5.2.1) or one union only represents workers of one skill group, each earning the same wage (as in Chapters 3.1.2 and 6.3.1).

The final critical assumption needed in order for the objective function (3.1) to be valid, is that all people employed by the firm are in fact also members of the union.[7] Only in this case is it possible for the union to prevent non-unionised workers from undermining the bargained wage. Although such a "closed shop" is forbidden in most countries, it was shown in Table 3.1 that the negotiated wage applies to far more workers than are actually union members. Therefore, even if a firm is able to hire a non-unionised worker, it is not guaranteed that it can employ him at a lower wage than unionised workers. Further, in Germany for example, under certain conditions laid out in §5 of the tariff negotiation law (the *Tarifvertragsgesetz* – TVG), the German Labour Minister has according to the *"Allgemeinverbindlicherklärung"* the right to declare the negotiated wage to be generally binding.[8] Finally, as unions have the power

---

[6] The median voter model was first formulated by Black (1948) and Arrow (1950). See Booth (1984) for a median-voter model within a unionised setting.

[7] So-called "insider-outsider" models which go back to seminal work by Blanchard and Summers (1986) and Lindbeck and Snower (1989) analyse the case where not all workers are represented by a union. This assumption also guarantees that no "free-rider" problem can arise whereby a worker enjoys the benefits of being in a unionised firm but does not have to become a member of the union and pay a membership fee in order to reap these benefits. For models which analyse such membership decisions, see Booth (1985) or Bulkley and Myles (2001). For an overview of "insider-outsider" models in general see Lindbeck and Snower (2001). However, as the distinction between insiders and outsiders is to a certain extent arbitrary, e.g. which group do part-time workers belong to, these theories will not be pursued here in any more detail.

[8] In Germany wage negotiations take place at the industry level. Thus, the Minister can only declare wages to be binding within a certain industry. In order to be able to do this, either the unions or the employers must submit a request and at least 50% of the workforce in the industry must be covered by the wage contract. Further, a representative of the top union as well as of the top employer federation must also give their consent. With the exception of the building trade, such declarations have only seldom been put into practice. See, for example, Franz (2003, Chap. 7) for a more detailed discussion concerning union laws in Germany.

to push through strikes and the firm may face large hiring and training costs when employing new workers, firms will also be reluctant to hire workers at non-union conditions.

For the case that there are more union members L_U than find employment, i.e. $L_U > L$, and union members do not possess any assets so that they are forced to find employment elsewhere and earn an alternative wage given by $\bar{w}$, then the utility function as given by equation (3.1) becomes

$$V = L(w)\nu(w) + (L_U - L(w))\nu(\bar{w}) \tag{3.2}$$

Assuming that the number of members is exogenous means that the utility maximum is given by

$$\frac{\partial V}{\partial w} = \frac{\partial L}{\partial w}[\nu(w) - \nu(\bar{w})] + L(w)\frac{\partial \nu}{\partial w} \overset{!}{=} 0$$

Thus, as only employed workers receive the utility-maximising wage, it is this number which is the binding restriction and only these workers will be considered by the union. Therefore, equation (3.1) can be modified to

$$V = L(w)[\nu(w) - \nu(\bar{w})] \tag{3.3}$$

An alternative to the above utilitarian concept, is to model union utility according to a Stone–Geary function. In this case, union utility is given by

$$V = (w - \bar{w})^\vartheta (L - L_0)^{1-\vartheta} \tag{3.4}$$

where L_0 is the minimum employment level that unions will tolerate and ϑ measures the degree of importance that unions place on wages as opposed to employment. In this case, the utility of the individual workers no longer features as an argument of the union welfare function. Therefore, this formulation of the utility function is more appropriate if one assumes that the union is primarily concerned with pushing through the interests of the union leaders.

The above analysis assumes that all union members have identical preferences. Even if this a simplifying assumption, which utility concept is chosen is at least partly a normative question. However, seeing as union leaders need to be reelected, they must try and represent the interests of their members. For this reason and in line with most of the literature, the utilitarian concept as given by equation (3.3) will be applied throughout.[9]

The reasons why only wages and employment feature in the union utility function have been specified above. However, just as important when analysing union behaviour is whether unions and employers bargain over wages

---

[9] See Booth (1995, Chap. 4) for a survey of empirical evaluations of union objective functions.

only, or whether bargaining covers both wage and employment aspects. The former are analysed in *"monopoly union"* models, in which case the union determines the utility-maximising wage given the labour-demand function, or in *"right-to-manage"* models where firms and unions bargain over the wage rate and firms subsequently determine labour demand as a function of the negotiated wage rate. In models of *"efficient bargaining"*, bargaining takes place over both wages and employment simultaneously. McDonald and Solow (1981) were the first to show that firms and unions could achieve Pareto superior results, i.e. at least one party could be made better off without making the other worse off, if the bargaining contract also specified the employment level. However, as shown by Layard and Nickell (1990), this result only holds in a partial equilibrium setting. In a general equilibrium, employment can actually decrease as a result of bargaining over employment. The economic intuition here is that such bargaining effectively increases union bargaining power as they are no longer faced with an invariant labour demand curve. It is precisely this last effect which seems able to explain the empirical fact that there is little evidence (see Oswald 1993) of union contracts covering both wages and employment. Further, as shown for example in Layard et al. (1991), even if bargaining is over wages and *effort*, that this too will fail to increase employment. It is for this reason that the subsequent analysis will only take bargaining over wages into account. In addition, seeing as it is extremely unrealistic that unions have all the bargaining power as assumed in the "monopoly union" model and further, as such a model can be treated as a special case of the "right-to-manage" setting, the models presented below will only focus on this latter type of bargaining.

The first important microeconomically founded paper analysing union behaviour and employing a utilitarian concept is by Oswald (1982).[10] In order to best be able to understand the working of the model, this primary analysis will make several simplifying assumptions in order to highlight the effects on the wage and employment levels of unions when wage bargaining takes place in one sector of the economy. Thus, in contrast to Oswald (1982), risk-neutral workers will be assumed here and the capital stock is treated as a constant equal to unity. Further, aspects such as the union influence on wage differentials between unionised and non-unionised workers will be left until Chapter 4.2.1 and the influence on growth will be analysed in Chapters 5.2.1 and 6.3.1.

3.1.1 Homogeneous Workers

For the reasons stated above, a union directly represents a fraction of the workforce in a particular firm, with the non-unionised workers employed un-

---

[10] Ever since the 1980s, the literature on unions has grown enormously, so that only the most important developments can be analysed here. For a more complete survey of the literature on union wage bargaining see Oswald (1985) or Pencavel (1994).

der the same conditions. This amalgamation of workers means that unions
have a monopoly with respect to labour supply and hence have more bargaining power than an individual worker acting on his own has. However, this is
only one condition which is necessary for unions to be able to operate effectively. The second required condition is the existence of a surplus or economic
rent in the sector in which the firm is operating as only in this case will it
be possible for competitive firms to pay above market-clearing wages without making losses. If, as assumed in this chapter, the firm faces a perfectly
competitive product market, then profits can occur for two different reasons.
Firstly, the production technology is such that it exhibits increasing returns
to scale at low employment levels and then decreasing returns as soon as employment is higher than a certain threshold value. This is illustrated in Fig.
3.1 which shows how firm revenue as a function of the price p and the production technology $F(L)$, which is assumed to be a function with only labour
as variable input, and labour costs fluctuate with different levels of labour
input. The competitive wage rate paid in a perfectly flexible labour market
is denoted by w_c. Alternatively, the production technology of the firm being

Fig. 3.1: Profits in a Competitive Industry

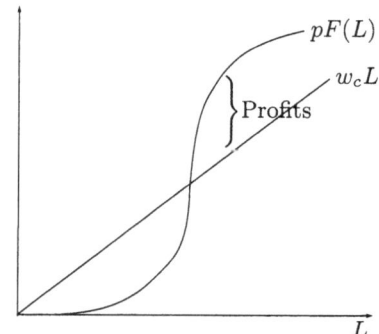

analysed is more advanced than that of its direct competitors so that it can
pay a wage above its marginal costs without losing all of its market share.

Assuming such production technologies and thus the existence of economic
rents, unions and employers are dependent on each other and can do better
by cooperating rather than acting independently. Hence, a bargaining problem arises between bilateral monopolists, with the firm as the single buyer of
labour confronted by a union as the single seller. The outcome of the bargaining process determines how the gains from the cooperation are divided. As
outlined above, bargaining is assumed to be of the "right-to-manage" type, i.e.
unions and employers jointly bargain only over wages with employers unilaterally setting employment levels according to their labour demand function as
derived from the production technology. This type of bargaining is almost al-

ways modelled using the bargaining concept developed by Nash (1950).[11] The underlying assumption behind this concept is that both bargaining parties want to maximise the joint surplus which is denoted by the Nash product or maximand. This function is the product of the objective functions of both parties weighted by their respective bargaining power. For the reasons stated above, the utilitarian concept is employed here so that the union objective function is the sum of the utilities of the individual union members. For the profit-maximising firm, their objective function is identical to their profit function. Further, for both parties there is a certain *"fallback position"* below which it is no longer optimal to continue bargaining in which case no agreement is reached. For the unions, this fallback position is reached when union utility is zero, which, as can be seen from equation (3.3), occurs when the bargained wage is identical to the outside option that unionised workers face. This outside option can be interpreted as the reservation wage $\bar{w}$ that workers could obtain in another sector of the economy or guaranteed income in the form of unemployment benefits. Thus, if the bargained wage were to drop below this level, a worker would no longer have an incentive to remain with the firm.[12] Of course, before this fallback position is reached, the union could also threaten to strike in which case the firm would also lose all output and thus profits.[13] Similarly, if firm profits were to become negative, the firm would be better off by not producing and so would pull out of the bargaining process so that the firm's fallback position is given by its zero-profit condition. Even if these fallback positions are not reached does not mean that they are unimportant: it is the *threat* of these fallback positions which counts and, as will be shown below, to the extent that the bargained wage is defined relative to the value of the outside option, the fallback position also influences the absolute wage level.

Normalising the price level in the sector which is being analysed to one, and with labour as the only factor of production, means that the firm's profits can be written as

$$\pi = F(L) - wL \tag{3.5}$$

where $F(L)$ is the firm's production function which is assumed to have decreasing returns in the vicinity of the equilibrium. Thus, from the above, the

---

[11] The alternative approach to such bargaining problems is a game-theoretic approach whereby the bargaining *process* is modelled with each side making alternative offers – see, for example, Rubinstein (1982), Binmore et al. (1986) or Sutton (1986). However, as shown in Booth (1995, Appendix 5A), under certain conditions, these two approaches yield identical results.

[12] If union members incur membership costs, then the fallback wage will be higher than the outside option by the amount necessary to cover these costs.

[13] Seeing as strikes are nowadays only seldom observed, this threat will not be analysed here. Instead, the interested reader is referred to Kennan (1986) for a more detailed discussion of strikes.

Nash product Ω is derived as

$$\max_{w} \Omega = \left[L[\nu(w) - \nu(\bar{w})]\right]^{\beta}\left[F(L) - wL\right]^{1-\beta} \qquad (3.6)$$

where $\beta \in [0,1]$ denotes union bargaining power. This parameter not only depends on the union's ability to act as a monopolist in the supply of labour, but also on institutional settings. For example, only if strikes are legally allowed and the union can convincingly state that it has the funds to finance such a strike, is the threat of a strike credible. Similarly, if there are restrictive firing rules then this is likely to strengthen the bargaining position of the unions as these limit the possibilities of employers to react to wage increases. Further, the degree to which a firm can legally circumnavigate union wages by hiring non-unionised workers who receive a lower wage will also be of influence.

If $\beta = 1$, all the bargaining strength lies with the union. This case corresponds to the "monopoly union" setting and leads to a wage rate that maximises union utility subject to the labour demand function. At the other extreme, for $\beta = 0$, firms will maximise profits under the condition that wages cannot fall below the worker reservation wage $\bar{w}$. If this wage is identical to the wage in a different, perfectly competitive, labour market, then the firm can push the wage down to its market-clearing level in this sector as well.

Maximising the Nash product as given by equation (3.6) with respect to the wage rate yields

$$\frac{d\Omega}{dw} = \beta\left[L[\nu(w)-\nu(\bar{w})]\right]^{\beta-1}\left[\frac{\partial L}{\partial w}[\nu(w) - \nu(\bar{w})] + \nu'(w)L\right]\left[F(L)-wL\right]^{1-\beta}$$
$$+ (1-\beta)\left[[\nu(w)-\nu(\bar{w})]L\right]^{\beta}\left[F(L) - wL\right]^{-\beta}\frac{\partial \pi}{\partial w} \overset{!}{=} 0$$

Since by the envelope theorem $d\pi/dw = \partial\pi/\partial w = -L$ holds, this simplifies to

$$\frac{\nu'(w)w}{\nu(w) - \nu(\bar{w})} = -\frac{w}{L}\frac{\partial L}{\partial w} + \frac{1-\beta}{\beta}\frac{wL}{F(L) - wL} \qquad (3.7)$$

Further, using the fact that

$$\frac{\partial \pi}{\partial L} = \frac{\partial F}{\partial L} - w \overset{!}{=} 0$$
$$\Rightarrow wL = \frac{\partial F}{\partial L}L$$

and the definition of the elasticity of output with respect to labour

$$\epsilon_{F,L} = \frac{\partial F(L)}{\partial L}\frac{L}{F(L)}$$

from which follows that

$$F(L) = \frac{wL}{\epsilon_{F,L}} \tag{3.8}$$

means that equation (3.7) can be rewritten as

$$\epsilon_{\nu,w} = \frac{1-\beta}{\beta} \frac{\epsilon_{F,L}}{1-\epsilon_{F,L}} - \epsilon_{L,w} \tag{3.9}$$

where $\epsilon_{\nu,w}$ and $\epsilon_{L,w}$ are the respective elasticities of utility and labour demand with respect to the wage rate. As long as their are decreasing returns to labour, $\epsilon_{F,L} < 1$ holds and the second fraction on the r.h.s of equation (3.9) is positive. The r.h.s. of equation (3.9) is a decreasing function of the bargaining power β. Thus, with higher levels of β, the elasticity of utility with respect to wages must also be falling. As the utility function is assumed to be concave, a lower elasticity corresponds to a higher wage level. Further, in absolute terms, the elasticity of utility with respect to the wage rate must exceed the elasticity of labour demand with respect to the wage rate. The reason for this is that although a wage increase directly increases the utility of those workers receiving this higher wage, there are two indirect counteracting effects. Firstly, seeing as the employers have the right to set employment, a higher wage will lower employment and secondly, higher wages will also decrease the firm's profits. In the limiting case with $\beta = 1$, unions will disregard this last effect as the employer's profit function has no direct influence on the bargained wage.[14] The bargaining result is depicted graphically in Fig. 3.2.

With union utility as a positive function of both wages and employment, the resulting indifference curves are convex to the origin.

Totally differentiating the profit function (3.5) for a given profit level yields

$$\frac{dw}{dL} = \frac{F'(L) - w}{L}$$

Whether the slope of this isoprofit line is positive or negative, depends on whether the marginal productivity of labour is smaller or greater than the wage rate. With decreasing returns to labour, the marginal productivity decreases with increasing levels of labour input. Therefore, the above function

---

[14] That unions are successful in creating a positive wage differential for their members has been empirically shown for the United States since Lewis (1963). Subsequently, there have been a vast number of empirical studies for other countries and for larger data sets confirming this result – see, for example, Kahn (1998) for recent evidence on several OECD countries and Farber (1986) or Lewis (1986) for surveys. Measuring the size of this wage differential naturally depends on a number of factors such as the chosen occupation or industry, and further on how *union* effects on wages are separated from other factors which influence wages, so that there is a large variation in estimates which range from 5 to 25 percent.

Fig. 3.2: Right-to-Manage Wage Bargaining

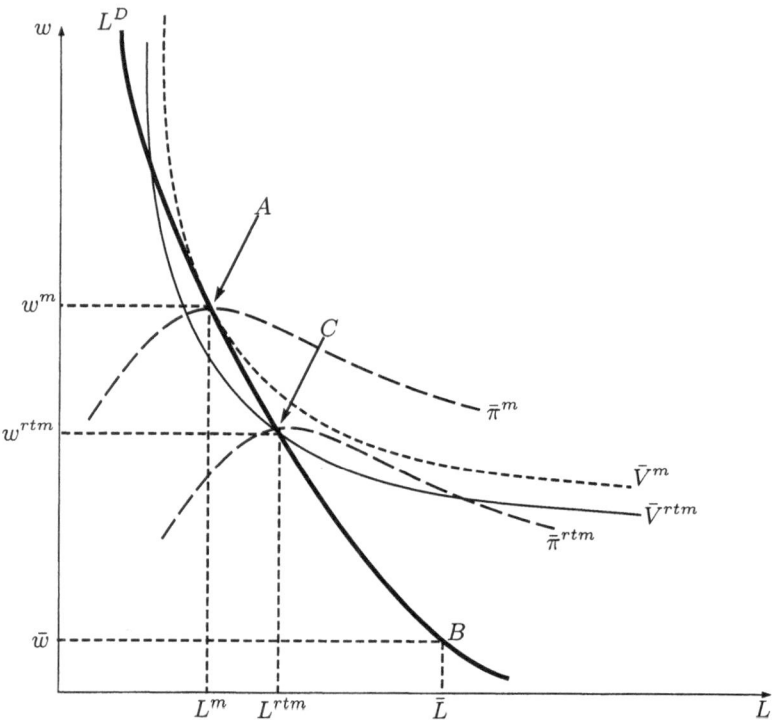

will first have a positive and then a negative slope. Optimal labour demand for the firm is where the marginal costs are exactly equal to the marginal benefits, so that at this point the maximum of the function is achieved. As shown in Fig. 3.2, the labour demand curve joins the maxima of all relevant isoprofit curves and defines a negative relationship between labour demand and wages.

The lines $\bar{\pi}^m$ and $\bar{\pi}^{rtm}$ denote isoprofits associated with the monopoly union and right-to-manage outcome respectively. Seeing as higher lying isoprofit lines are associated with higher wages (for a given employment level), lower isoprofit curves imply a higher profit levels. Therefore, point A in Fig. 3.2 denotes the outcome when the union has all the bargaining power and acts as a monopoly union. In this case, workers receive the wage w^m, firms earn profits π^m and L^m workers find employment. Firms earn the highest profits if they have all the bargaining power, i.e. $\beta = 0$ and can set the wage equal to the reservation wage $\bar{w}$, which here is assumed to be identical to the fully competitive, market-clearing level (point B in Fig. 3.2). Of course, if unions have no bargaining power, the labour market is in fact perfectly competitive so that full employment is reached at point B, where $\bar{L}$ denotes total labour supply at the competitive wage level. Thus, points A and B mark the maximum and minimum wage outcomes and the minimum and maximum employment levels. For all intermediate values of the bargaining, a wage rate in between these two extremes will result with the corresponding employment level given by the labour demand function. Point C denotes such an outcome. Thus, as soon as market imperfections such as union wage bargaining are introduced, it is no longer possible to reach the first-best solution. Thus, higher union bargaining strength leads to higher wages but at the expense of a labour market which is not fully cleared. As these higher wages also lead to firms earning lower profits, unions also have a redistributive effect. Whether this latter effect is desired or not, is a normative question. However, this redistribution undoubtedly comes with the cost of a loss in efficiency so that the question arises whether there are not more efficient means (e.g. the tax system) of reaching the same redistributive goal.

Although leaving a more general critique of union wage bargaining models until the end of the next section, one of the weak points in the analysis so far is clearly that labour is treated as homogeneous, an assumption which is clearly at odds with reality. This assumption is relaxed in the next section.

3.1.2 Heterogeneous Workers

As shown in Chapter 2, starting in the late 70s, a large increase in the supply of skilled labour has been observed in all developed economies and the nature of technological change has been biased towards high-skilled labour. Although a detailed discussion of the theoretical explanations for this skilled-biased technological change are left until Chapter 6, it is clear that modern theoretical models must differentiate between low- and high-skilled labour.[15]

The analysis here and throughout will assume that the distribution of these skills is exogenously given, i.e. a constant fraction of workers are assumed to be "high-skilled", and the remainder low-skilled. Thus, why some workers become highly skilled and others do not remains outside of the analysis. However, by assuming that workers have different learning abilities and thus different learning "costs", it is easily possible to endogenise the fraction of workers who become highly skilled – see, for example, McConnell et al. (1999, Chap. 4). However, with such a setup, no additional insights are gained, as there is still only one (exogenous) parameter that explains who acquires how many skills. Further, it is assumed that it is perfectly observable whether a worker is of a high- or low-skill type. Thus, we abstract from information asymmetries as pointed out by Spence (1973), whereby a worker first hast to acquire a certain educational degree in order to credibly signal the potential employer that he is highly skilled.

---

[15] In almost all of the literature which differentiates labour by skill, labour is divided into only these two categories. However, see, for example, Fitzenberger (1999) where labour is divided into finer skill categories.

Although Barth and Zweimüller (1995) are mainly interested in comparing decentralised and corporatist bargaining systems, they disregard all employment consequences of wage bargaining by assuming that union utility is only a function of wages. This restrictive and unrealistic assumption is dropped making it possible to extend the model from above to the case where labour is heterogeneous with respect to skills.[16] To simplify the analysis, it is assumed that each type of labour is represented by a separate union and that the two unions do not coordinate their actions. Further, the two types of labour are assumed to be gross complements as inputs in production, i.e. demand for labour of one skill type will decrease if the wage rate for the other skill type increases. Thus, although the two types of labour are imperfect substitutes for each other, both are needed for production, so that a strike by one labour group automatically halts all production. Normalising the price level in this sector to unity means that the profit function entering the Nash product when the employer bargains with the union representing low-skilled labour is now

$$\pi - \pi_0 = F[L_L(w_L, w_H), L_H(w_L, w_H)] - w_L L_L(w_L, w_H) - w_H L_H(w_L, w_H)$$
$$- [-w_H L_H(w_L, w_H)]$$
$$= F[L_L(w_L, w_H), L_H(w_L, w_H)] - w_L L_L(w_L, w_H) \qquad (3.10)$$

where π_0 is the firm's fallback position and an index L denotes low-skilled and H high-skilled labour, respectively. An analogous equation holds when the firm bargains with the union representing high-skilled workers.

Assuming the same union utility function as given by equation (3.3), the Nash product when the employer bargains with the union representing the low-skilled workers is now given by

$$\max_{w_L} \Omega = \left[L_L[\nu(w_L) - \nu(\bar{w}_L)]\right]^{\beta_L} \left[F[L_L(w_L, w_H), L_H(w_L, w_H)]\right.$$
$$\left. - w_L L_L(w_L, w_H)\right]^{1-\beta_L}$$

with $\bar{w}_L$ denoting the wage a low-skilled employee could obtain in a perfectly competitive non-unionised sector of the economy, and β_L as the bargaining

---

[16] Olson (1984) was among the first to analyse the effects of different degrees of centralisation of union bargaining on the economic outcome. His model predicts an inverted U-shaped relationship between the unemployment rate and the degree of centralisation. The reason for this is that with fully decentralised bargaining, the union takes into account that any wage increase cannot be passed on to consumers through higher product prices but will instead lead to a decrease in employment. Similarly, with fully centralised bargaining, all negative effects of wage bargaining are internalised, so that at these two extremes, the economic outcomes are the same, whereas employment is lowest when bargaining takes place at the industry level. This result was empirically confirmed by Calmfors and Driffill (1988). Layard et al. (1991, Chap. 4, Table 4) provide empirical evidence for OECD countries showing that the degree of responsiveness to firm-specific shocks is larger if bargaining is at the firm level.

power of the union representing the low-skilled workforce. Maximising the above expression with respect to the low-skilled wage rate and writing the production function without its arguments for notational convenience yields

$$\frac{\partial \Omega}{\partial w_L} =$$

$$\beta_L \left[L_L(\nu(w_L) - \nu(\bar{w}_L)) \right]^{\beta_L - 1} \left[\frac{\partial L_L}{\partial w_L} [\nu(w_L) - \nu(\bar{w}_L)] + L_L \nu'(w_L) \right]$$

$$\left[F - w_L L_L(w_L, w_H) \right]^{1 - \beta_L} + \frac{\partial F}{\partial L_L} \frac{\partial L_L}{\partial w_L} + \frac{\partial F}{\partial L_H} \frac{\partial L_H}{\partial w_L} - L_L - w_L \frac{\partial L_L}{\partial w_L} \overset{!}{=} 0$$

Noting that $\partial F / \partial L_L = w_L$, the above simplifies to

$$\frac{\frac{\partial L_L}{\partial w_L}[\nu(w_L) - \nu(\bar{w}_L)] + L_L \nu'(w)}{L_L[\nu(w_L) - \nu(\bar{w}_L)]} = \frac{1 - \beta_L}{\beta_L} \frac{L_L - w_H \frac{\partial L_H}{\partial w_L}}{F - w_L L_L}$$

$$\frac{\nu'(w_L) w_L}{\nu(w_L) - \nu(\bar{w}_L)} = \frac{1 - \beta_L}{\beta_L} \frac{w_L L_L - w_L w_H \frac{\partial L_H}{\partial w_L}}{F - w_L L_L} - \epsilon_{L_L, w_L} \tag{3.11}$$

where ϵ_{L_L, w_L} denotes the elasticity of low-skilled labour demand with respect to the low-skilled wage rate. Equation (3.11) can be further simplified to

$$\epsilon_{\nu, w_L} = \frac{1 - \beta_L}{\beta_L} \frac{\epsilon_{F, L_L} - \omega}{1 - \epsilon_{F, L_L}} - \epsilon_{L_L, w_L} \tag{3.12}$$

Here, ϵ_{ν, w_L} is the elasticity of utility with respect to the low-skilled wage rate and ϵ_{F, L_L} is the output elasticity with respect to low-skilled labour input. The term $\omega \equiv w_L w_H (\partial L_H / \partial w_L) / R$ is a constant where R denotes revenue. Note that ω is equal for both groups as in equilibrium (with profit maximisation), the substitution matrix of factor inputs is symmetric (see, for example, Varian 1992, p. 34). Further, with the assumption that the two types of labour are gross complements in production, $\omega < 0$. Therefore, a comparison of equation (3.9), which describes the case of homogeneous labour, and equation (3.12) which describes the bargaining solution for heterogeneous labour, immediately shows that the r.h.s. of equation (3.12) is higher which, with a concave utility function, implies that the bargained wage rate is lower than in the case of homogeneous labour. The economic intuition for this result is that now not only do unions have to take the negative effects of higher wages on labour of their skill type and on firms' profits into account, but in addition, a higher wage for one skill group lowers the demand for labour of the other skill group as well. This means that the productivity of the labour type represented by the union declines which further reduces demand for labour of this skill type.

The equation for the high-skilled wage rate is derived analogously. Even if the wage rates for the two skill types are above their market-clearing levels, whether the wage differential between the two groups is higher or lower than it would be without unions depends on the precise functional form of the union utility functions, the production function and on the relative bargaining strengths of the two respective unions. Whether the bargaining power of the union representing the low-skilled workers is higher or lower than the corresponding value for the union representing the high-skilled is an empirical question, so that its effect on the wage differential cannot be determined theoretically. However, for the empirically plausible case that the union for the low-skilled has the greater bargaining strength, implies that the wage differential will be lower than in a fully competitive labour market which corresponds to what is observed internationally, see, for example, Blau and Kahn (1996) for evidence.

As the high-skilled are more productive, their wage rate will be higher than that for low-skilled workers. Assuming that both the low- and high-skilled workers have identical (concave) utility functions means that with union bargaining, the l.h.s. of (3.12) will be lower for high-skilled workers. Further, the higher productivity of the high-skilled also means that their output elasticity ϵ_{F,L_H} is higher. Hence, it can be seen from equation (3.12) that (in absolute terms) the elasticity of labour demand with respect to wages must be higher for the low-skilled than the corresponding value for the high-skilled. Again, this is in accordance with the empirical evidence (see Fitzenberger 1999).

With the presence of unions as the sole cause of labour-market imperfections, whether the unemployment rate is higher or lower for the low-skilled only depends on the relative bargaining strengths of the unions representing the two types of workers. However, assuming as above that the union of the low-skilled has the higher bargaining strength also implies that their unemployment rate will be higher as their wage rate will be a higher markup over the market-clearing level.

The effects of an increase in the supply of highly skilled labour also depend on which union has more bargaining power. Assuming isoelastic functions means that the elasticities given in equation (3.12) are independent of the absolute values of labour supply so that an increase in the supply of high-skilled labour will c.p. not lead to a change in the unemployment rate for either skill group. Therefore, whether the unemployment rate rises or falls with such a shift in labour supply depends on whether the size of the group with the lower skill-specific unemployment rate increases or not. However, it should be noted that these conclusions are based on a partial-equilibrium analysis. It is unlikely though, that such a shift in labour supply is restricted to one sector of the economy. For this reason, a more detailed analysis requires a general-equilibrium model so that further discussion of this aspect will be left until Chapter 6.3.1 where the interaction of innovation-based growth and unions is analysed.

3.1.3 Discussion

This chapter so far has shown that the presence of unions in a labour market, irrespective of whether labour is treated as homogeneous or not, leads to a wage rate above its competitive level. These higher wages result as unions, with their monopoly power over the supply of labour, have more bargaining power than an individual worker and are therefore able to force employers to share any rents that accrue on the market by accepting lower profits and instead paying a higher wage. This leads to a conflict between equality and efficiency as the "price" of these higher wages is an allocative inefficiency in the form of unemployment. However, it should be noted that this result implicitly assumes a perfectly competitive labour market as a benchmark, an assumption which is only justifiable on theoretical grounds. Thus, if other labour-market imperfections were to be included, for example efficiency-wage considerations discussed in the next section in more detail, a higher union wage may be beneficial for the firm which faces a trade-off between higher wages and higher monitoring costs.[17] Similarly, as shown for example in Eguchi (2002), if unions have the effect of making it more difficult for firms to dismiss their workers, then both firms have a larger incentive to invest in the human capital of their workforce as well as workers themselves will be prepared to accept lower wages in return for more job security. Further, the above results were proven under the assumption of a perfectly competitive product market and perfect information. Whilst this makes it possible to focus entirely on the wage and employment effects caused by unions, it also means that potentially efficiency-enhancing effects such as lower negotiation costs before a wage settlement is reached for each worker or better communication between employers and employees leading to higher productivity levels are neglected.

One of the main assumptions implicitly made in the union models above, is that the employer is interested in paying the lowest wage possible as any increase in the wage rate automatically leads to a reduction in profits. However, such an assumption neglects any positive effects higher wages may have on productivity. It is these effects which are the focus of the efficiency-wage theories which are analysed in more detail in the next section.

3.2 Efficiency Wages

As shown in Chapter 2, ever since the 1970s, the wage rate has remained remarkably stable or actually risen even in times of high unemployment. In the union wage-bargaining models discussed above, it was assumed that collective

---

[17] Brown and Medoff (1978) provide empirical evidence in support of this argument. See Blank and Freeman (1994) for a discussion of how unions can correct for labour-market failures.

agreements between unions and employers prevent wages from falling. However, the conclusion that unions and the distortions they create on the labour market are the sole cause of unemployment is empirically false. As Layard et al. (1991, p. 150) put it: "Unions are a newer phenomenon in human history than unemployment". As will be shown in more detail below, if paying higher wages also leads to higher productivity, then firms also have an incentive to increase the wage they pay. In this context, the costs of paying a higher wage must be weighted against the gains in output associated with this wage increase. It is these beneficial, productivity-enhancing effects of higher wages that are the focus of efficiency-wage theories. Although most efficiency-wage models were not developed until the late 70s and early 80s, Henry Ford was probably the first employer to use the concept when in 1914 he doubled the minimum wage paid at Ford manufacturing plants and later said: "This was one of the finest cost cutting moves we ever made" (Raff and Summers 1987). The benefits of higher wages vary depending on the type of efficiency-wage model under consideration.[18] However, all models have in common that the output of a worker is a positive function of the amount of "effort" a worker is prepared to provide. There are several reasons for assuming that productivity and hence effort are positively linked to the wage the employer is prepared to pay.[19]

Firstly, before production even starts, by announcing a higher wage, a firm will be able to attract applications from more highly qualified workers with relatively high reservation wages. This idea, which goes back to Weiss (1980), helps firms reduce the problems of adverse selection as through their wage decisions, firms can signal which kind of applicants they are searching for.[20] Hence, assuming that the qualifications of a worker are positively correlated to his productivity as is realistic, such a measure will lead to a more productive workforce and therefore, c.p. to higher output and revenue levels for the firm.

Secondly, in the so-called "*labour-turnover*" models which go back to Schlicht (1978) and Salop (1979), a higher wage will reduce labour turnover amongst the workforce as workers are less likely to quit the higher their wage in the current job is. Therefore, fewer new workers need to be hired to replace workers who have left the firm. This has the effect of lowering training costs which in turn raises average productivity as a smaller fraction of the current employees

---

[18] For detailed surveys of different efficiency-wage models, see, for example, Yellen (1984), Akerlof and Yellen (1986) and Katz (1986).

[19] The reasons stated in the following all apply to industrialised economies. In fact, the first efficiency-wage model goes back to Leibenstein (1957) who argued that if a firm in a developing country paid a higher wage, then the workforce would be able to buy itself better food and therefore be physically stronger and healthier and therefore more productive.

[20] See, for example, Holzer et al. (1991) for evidence.

are involved in training new employees, an activity which hinders them from fully carrying out their main task in the firm.[21]

Thirdly, there is the *"no-shirking"* class of efficiency-wage models first developed by Shapiro and Stiglitz (1984). Here, firms are assumed to be unable to perfectly monitor their workforce and hence face a trade-off between either spending more on a costly monitoring technology or paying their workforce a higher wage. If it chooses to do the latter, the higher wage acts as a disciplinary device to prevent shirking. The reason for this is that if the wage the firm were to pay is the competitive wage so that the labour market fully clears, and any worker detected shirking is immediately dismissed, then this laid-off worker would immediately be able to find a new job at the same wage in a different firm. Therefore, being dismissed for shirking is not associated with a loss in income for the worker who thus has no incentive to provide effort. However, if the wage the firm pays is above the competitive level, then a worker must take into account that if he is caught shirking and dismissed, he will not be able to receive the same wage elsewhere. Therefore, although the worker has a cost of providing effort to the firm, he also has a benefit from the higher wage which he cannot earn elsewhere. Hence, now the threat of immediate dismissal for shirking is much more effective as a worker who loses his job will also suffer a reduction in his earnings and hence utility. Of course, if it is rational for one firm to pay a wage above the market-clearing level, it is rational for all firms to handle accordingly so that now unemployment will occur. In this case, even if a dismissed worker will with a certain probability find a new job paying the same wage as his previous company, he will suffer from an unemployment spell with reduced earnings before being able to find this new job so that dismissal still involves an income reduction.[22]

The fourth and final major category are the *"fair-wage"* models. This class of models stems from psychological and sociological observations on the interaction between employers and employees. These include production and working conditions, as well as social norms and values. Further, attitudes prevailing within a firm can have a decisive influence on how conflicts are managed, on the hierarchical structure of a firm and hence on the motivation of workers. Essentially, if workers feel themselves "fairly" treated by their employers, they will in return be prepared to provide effort at work. It is for this reason, that these models which date back to seminal work by Akerlof (1982, 1984) and Akerlof and Yellen (1986, 1988, 1990) are often stated to be a type of "gift-exchange". Of course, what precisely is meant by fair is a normative question, but in economic terms, the models assume that fairness is judged by either comparing one's own wage with that of other workers in the same firm or, if possible, one's own wage with the wage paid to a similarly qualified worker

---

[21] See Campbell (1993) for empirical evidence.

[22] See, for example, Cappelli and Chauvin (1991), Rebitzer (1995) and Ewing and Payne (1999) for empirical evidence of higher wages leading to lower shirking levels.

employed in a different firm.[23] Therefore, if he finds that he is earning relatively too little compared to other workers, then he will reduce his supply of effort accordingly. There is mixed evidence as to how workers react if they feel themselves "overpaid".

As stated above, all efficiency-wage models have many elements in common irrespective of how precisely the productivity-enhancing effects of higher wages occur. However, in the sense that if a worker feels himself fairly treated he will also have fewer incentives to shirk or leave the firm, efficiency-wage models of this category can be interpreted as encompassing all other efficiency-wage categories as special cases. It is for this reason, that this most general type of efficiency-wage model is discussed below, first for the case of homogeneous workers, and subsequently when labour is heterogeneous.

3.2.1 Homogeneous Workers

As briefly described above, the starting point for efficiency-wage theory is that workers have a certain amount of control about how much effort they are willing to provide. However, now utility not only (positively) depends upon the (relative) wage the worker receives, but also on the disutility of exerting effort $\tilde{e}$. Hence, utility ν of a representative agent is given by

$$\nu - \nu(w, \tilde{e}), \qquad \frac{\partial \nu}{\partial w} > 0, \frac{\partial \nu}{\partial e} < 0 \tag{3.13}$$

From the firm's viewpoint, the central concept behind all efficiency-wage theories is that a higher wage paid to a worker will induce him to provide a higher amount of effort which leads to higher output levels and hence an increase in profits and lower unit-labour costs. In order to focus on the labour-market imperfections solely caused by paying efficiency wages, it is assumed that there are no unions present in this labour market. Hence, an employer can unilaterally optimally set both the wage as well as the employment level. With productivity positively depending on the (relative) wage rate, the production function is now assumed to be given by $F(\tilde{e}(w, \bar{w})L)$, with $\partial \tilde{e}/\partial w > 0, \partial \tilde{e}/\partial \bar{w} < 0$ and $\partial F/\partial(\tilde{e}(w, \bar{w})L) > 0$ and $\partial^2 F/\partial(\tilde{e}(w, \bar{w})L)^2 < 0$, where $\bar{w}$ is again assumed to be the market-clearing wage which any worker could obtain in a labour market operating in a different sector of the economy. Maximisation of profits leads to

$$\pi = F(\tilde{e}(w, \bar{w})L) - wL \tag{3.14}$$

which leads to the following first-order conditions:

---

[23] This equity theory was first formalised by Adams (1963). See, for example, Summers (1988), Akerlof and Yellen (1990) or Gächter and Falk (2002) for empirical support of the above assumptions.

$$\frac{\partial \pi}{\partial L} = \frac{\partial F}{\partial (\bar{e} L)} \bar{e}(w) - w \overset{!}{=} 0 \tag{3.15}$$

$$\frac{\partial \pi}{\partial w} = \frac{\partial F}{\partial (\bar{e} L)} \frac{\partial \bar{e}}{\partial w} L - L \overset{!}{=} 0 \tag{3.16}$$

Combining these two equations yields the so-called "Solow" condition

$$\frac{\partial \bar{e}'(w) w}{\bar{e}(w)} = 1 \tag{3.17}$$

which dates back to Solow (1979). This condition states that in the optimum, the marginal revenue gained by an increase in the wage due to the higher effort and hence productivity level, must be equal to the additional marginal cost of the higher wage, i.e. the optimum is characterised by minimal wage costs per efficiency unit. Figure 3.3 depicts this result graphically. The effort-supply

Fig. 3.3: The Determination of the Efficiency Wage

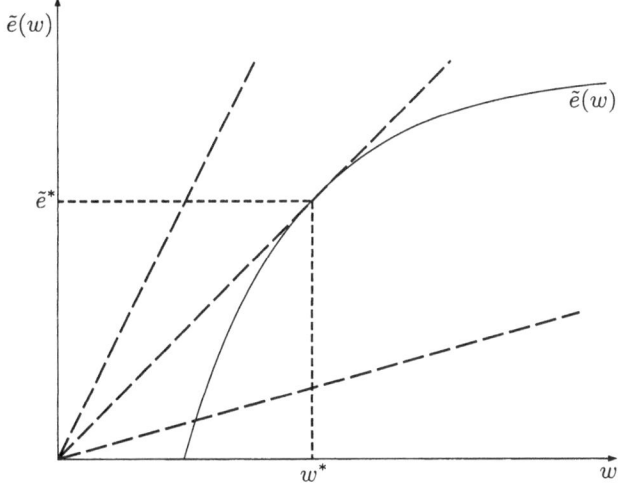

function with respect to wages is assumed to be concave. It does not start in the origin as it is assumed that a minimum wage must be paid before workers are prepared to supply any effort.[24] How high this minimum wage is, depends on the specification of the utility function (3.13). The rays coming out from the origin are lines with constant effort to wage ratios. The firm naturally prefers higher rays as, for a given wage level, these denote a higher effort and therefore output level. The optimum is therefore characterised by the outermost ray which is just tangent to the effort function. In this point, with optimal effort $\bar{e}^*$ and optimal wages w^*, the marginal and average elasticity

---

[24] Alternatively, a unique efficiency wage results if it is assumed that the effort function starts in the origin but first exhibits increasing and then decreasing returns. See Goerke and Holler (1997, p. 219p.) for a more detailed discussion.

with respect to the wage are equal, which is exactly the optimal condition given by the Solow condition.

To show the implications of efficiency-wage setting more formally, the effort function is specified as

$$\tilde{e} = -\varepsilon_1 + \varepsilon_2 \left(\frac{w}{\bar{w}}\right)^{\varepsilon_3} \qquad \varepsilon_1 > \varepsilon_2(1 - \varepsilon_3), \ 0 < \varepsilon_3 < 1 \tag{3.18}$$

The Solow condition leads to an optimal effort level of

$$\tilde{e}^* = \frac{\varepsilon_1 \varepsilon_3}{1 - \varepsilon_3} \tag{3.19}$$

Inserting this value back into the effort function (3.18) yields

$$\frac{w}{\bar{w}} = \left(\frac{\varepsilon_1}{\varepsilon_2(1 - \varepsilon_3)}\right)^{\frac{1}{\varepsilon_3}} \tag{3.20}$$

From this relative wage and given the parameter restrictions stated above, it can be seen that this wage differential is greater than unity, i.e. that the wage paid in this sector will be above the market-clearing level. This wage differential is increasing in the negative intercept term ε_1 as c.p. an increase in this term leads to workers exerting less effort so that the only way the employer can counteract this effect, is by paying a higher wage. As a higher value of ε_2 leads to workers increasing the effort they provide, in this case the firm will lower the wage it pays until the Solow condition is again fulfilled.

Figure 3.4 depicts the equilibrium graphically. From the properties of the production function, equation (3.15) defines a labour-demand function which negatively depends on the wage. Similarly, given the assumptions about the production technology, equation (3.16) defines an upward sloping labour-supply function. In the absence of efficiency-wage considerations, the effort parameter $\tilde{e}$ is implicitly set to unity. However, as can be seen from the specification of the effort function as given by (3.18), due to the negative intercept term, the labour-supply curve is shifted upwards so that the wage rate now needs to be higher than the market-clearing level.

Although the efficiency-wage theory is of course subject to criticism, this will be left until the end of this section. For now, the above analysis will be extended to incorporate heterogeneous labour.

3.2.2 Heterogeneous Workers

The above analysis can easily be extended to include labour of different skill types. It is again assumed that the distribution of these skills is binary and

Fig. 3.4: Efficiency-Wage Setting

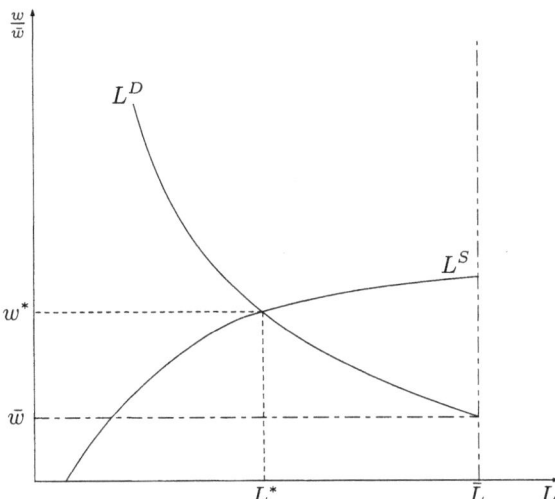

exogenously given and that they are perfectly observable so that information asymmetries are not taken into account.[25] If the two types of labour only make within group comparisons, then the above considerations will simply apply to both skill-types individually with one effort function specific to one type of labour, i.e. the functional form of the effort function (3.18) remains unchanged but the parameters need to be augmented by a skill index. If, however, the two types of labour also base their decisions on the wage rate which the other type of labour receives, then the above function needs to be modified, e.g. in the case of low-skilled workers to

$$\tilde{e}_L = -\varepsilon_{L1} + \varepsilon_{L2} \left(\frac{w_L}{\bar{w}_L w_H^{\varepsilon_4}} \right)^{\varepsilon_{L3}} \tag{3.21}$$

where $\tilde{e}_L$ is the effort supply of low-skilled workers and a corresponding effort-supply function holds for the high-skilled. From this equation and noting that the Solow condition must simultaneously hold for both skill groups means that the optimal effort levels are given by

$$\tilde{e}_s^* = \frac{\varepsilon_{s1}\varepsilon_{s3}}{1 - \varepsilon_{s3}}, \qquad s \in \{L, H\} \tag{3.22}$$

where s is again a skill index with L denoting low- and H high-skilled labour. Inserting these values back into the effort functions (3.21) results in

$$\frac{w_L}{\bar{w}_L} = w_H^{\varepsilon_4} \left(\frac{\varepsilon_{L1}}{\varepsilon_{L2}(1 - \varepsilon_{L3})} \right)^{\frac{1}{\varepsilon_{L3}}} \tag{3.23}$$

for the low-skilled and

---

[25] See Ma and Weiss (1993) for an efficiency-wage model with signalling.

$$\frac{w_H}{\bar{w}_H} = w_L^{\varepsilon_4} \left(\frac{\varepsilon_{H1}}{\varepsilon_{H2}(1 - \varepsilon_{H3})} \right)^{\frac{1}{\varepsilon_{H3}}} \tag{3.24}$$

for the high-skilled. From these two equations it can be seen that the low-skilled have a higher (sectoral) unemployment rate, i.e. the markup of their wage over the market-clearing wage $\bar{w}_L$ if

$$\varepsilon_{L1} > \varepsilon_{L2}(1 - \varepsilon_{L3}) > \varepsilon_{H1} > \varepsilon_{H2}(1 - \varepsilon_{H3}), \qquad 0 < \varepsilon_4 < 1$$

hold. This corresponds to the case where low-skilled workers place more emphasis on "fairness" considerations than their high-skilled counterparts. Further, in order to ensure a higher wage for the high-skilled

$$\frac{\bar{w}_H}{\bar{w}_L} > \frac{\hat{\varepsilon}_L}{\hat{\varepsilon}_H}$$

must also hold where $\hat{\varepsilon}_s$ is defined as

$$\hat{\varepsilon}_s \equiv \left(\frac{\varepsilon_{s1}}{\varepsilon_{s2}(1 - \varepsilon_{s3})} \right)^{\frac{1}{\varepsilon_{s3}}}$$

In the above analysis it has been shown that there are market imperfections on both the labour supply and demand side. On the one hand, workers are only willing to provide adequate effort if the wage is high enough to compensate them from the disutility of providing this effort. On the other hand, firms do not want the wage to fall to market-clearing levels as otherwise they cannot provide any incentives for their employees to produce any output. Nevertheless, the above model is difficult to reconcile with the empirical fact that the unemployment rate is lower for the high-skilled (see Chapter 2). If the two types of labour have identical utility and effort-supply functions, then the skill-specific wage differential between the efficiency wage and the market-clearing wage will also be identical (in relative terms). Hence, in this case, the unemployment rate for the two-types of labour will be the same. Thus, as the production technology is exogenous, only by arguing that, for example, the high-skilled are more motivated or place less emphasis on fairness considerations and therefore require a relatively lower wage in order to provide a given effort level, is it possible to generate a model where, c.p., the high-skilled are less likely to be unemployed than the low-skilled are.[26] Further, the model is also unable to explain the empirical fact, that an increase in the supply of high-skilled workers will not alter or even increase the wage differential between low- and high-skilled workers. Although the wage differential

---

[26] It is possible to assume that, due to exogenous shocks, there is a certain job-breakup rate for each job. For this case, Jones (1987b) has developed a model where it is assumed that the high-skilled are more flexible and can adapt to such shocks more easily so that firms prefer hiring these workers and hence that their unemployment rate is lower than that of the low-skilled.

between the high-skilled market-clearing wage and the high-skilled efficiency wage will remain unaltered (in relative terms), the increase in the number of high-skilled does mean that the market-clearing wage for this group must fall. However, with an unaltered wage markup for the high-skilled, such an increase in the fraction will also leave their unemployment rate unaltered, a fact which is in accordance with the empirical evidence. It needs to be noted though that these predictions are all based upon a partial-equilibrium analysis and the unemployment rates discussed in Chapter 2 of course all apply to the economy as a whole and not just one specific sector. For this reason, the results here need to be treated with caution and the question of skill-specific unemployment rates and the wage differential between the two types of labour will be analysed within a general-equilibrium setting in Chapter 6.3.2.

3.2.3 Discussion

As shown in Chapter 2, even in times of high unemployment, real wage levels did not fall to bring the labour market back into equilibrium. Whereas in Section 3.1 it was the presence of unions which prevented wage levels from falling, the efficiency-wage theory clearly shows that there are many reasons why firms also have an intrinsic interest in wages remaining above their market-clearing level. The above analysis concentrated on the fair-wage motivation for efficiency wages, but the same line of reasoning also applies if, for example, a firm can save costs by reducing its labour turnover rate or wants to reduce its monitoring costs and use efficiency wages as a disciplinary device to prevent workers from shirking.

However, even if the efficiency-wage theory can plausibly explain why the wage level remains above its market-clearing level, it is important to note that, as in equilibrium the wage is set so that all workers provide effort, the only explanation why people become unemployed is due to the exogenously given notion of fairness. Further, with unemployment occurring because the wage does not fall to market-clearing levels, the derivation of the wage rate is decisive. However, the wage level crucially depends on the functional form of the effort function which is exogenously given. Hence, the explanation of the wage and unemployment rate remains unsatisfactory. Further, as these exogenous factors will be extremely difficult for a government to influence, there are no potential policy implications which result.[27]

---

[27] In the simplest form of the no-shirking efficiency-wage models, the value of unemployment benefits does influence the unemployment rate, with higher benefits leading to a rise in unemployment as they require a correspondingly higher efficiency wage to ensure that workers have adequate incentives to provide effort. However, within such a model, Goerke (2000) shows that if unemployment benefits are higher for non-shirkers than for shirkers, that higher benefits can actually reduce unemployment. Further, the role that benefits play changes fundament-

Efficiency-wage models are often criticised in the sense that there are more effective methods than paying higher wages of extracting the required effort from individuals. In this context, it is argued that wages can be left at market-clearing levels whilst workers should be required to pay a bond when they are hired.[28] Hence, no unemployment would arise, but workers would still exert the required amount of effort. This bond is then paid back to the worker when he leaves the firm but is retained by the firm if the worker is detected shirking. However, such bonds are rarely observed. Apart from possible legal arguments against such a solution, it also has drawbacks from an economic point of view as now moral hazard problems arise. Clearly, under such a ruling, a firm has an incentive to (falsely) claim that the worker was in fact shirking in order to collect the bond. Further, in the absence of perfect capital markets, it can not be guaranteed that a worker at the beginning of his career has the funds to post a large enough bond.

In both the union bargaining as well as efficiency-wage models, unemployment is entirely explained by a wage rate that is above its market-clearing level. However, especially in times of fast structural change, it is important that a hired worker has the qualifications required to perform the job. Seeing as these qualifications are seldom universally met by all applicants, firms must invest time and money into finding a suitable worker. Similarly, a worker cannot apply for every available job, but instead must also find a suitable job offer. This exemplifies the fact that time and money needs to be spent to find a suitable job-worker pair, i.e. that labour market frictions play an important role. These are the focus of the next section.

3.3 Matching Processes

In the basic neoclassical labour market, all firms are identical and perfectly informed about all workers and vice-versa. Seeing as there is also perfect information about the location of all job offers and job seekers, all vacancies are instantaneously filled and all job-seekers immediately find a job. However, as anyone who has ever applied for a job or who has ever had to fill a position will know, the labour market is to a certain extent a search market with heterogeneities and imperfect information. In reality, firms need to allocate resources to search for workers with specific skills and needs and likewise, workers need to invest time, money and effort in order to find a job offer which, for example, requires the qualifications they have to offer. It is for

---

ally as soon as imperfect information (see Stiglitz 1986) is introduced or benefits are financed by labour as well as capital taxes or the wage-rental relationship is important for the effort supply (see Agell and Lundborg 1992).

[28] This was originally suggested by Yellen (1984) and Carmichael (1985). See Shapiro and Stiglitz (1985) and Akerlof and Yellen (1989) for a subsequent discussion.

this reason that models that analyse this economic problem are called search or in the sense that a productive job-worker pair is being brought together, "matching" models.[29]

The theory of search in the labour market goes back to the late 1960s – see, for example, Phelps (1968) or Alchian (1969) – when it became part of the general research being undertaken on the Phillips curve and the natural rate of unemployment. This early search theory assumed a distribution of wage offers for identical jobs with workers rejecting low-wage jobs. However, this version of the theory was rejected on both logical (Rothschild 1973) and empirical grounds (Tobin 1972 and Barron 1975). The first model to overcome these criticisms was by Pissarides (1979), who applied a zero-profit condition for new jobs to endogenise the demand for labour. Although many modifications of the theory have been made since then, it is this model as presented in Pissarides (2000, Chap. 1) which builds the basis for the analysis below.

The heterogeneities and information imperfections mentioned above lead to frictions in the labour market and therefore to the coexistence of job openings and unemployment – which are negatively correlated as shown in the Beveridge curves as explained in Chapter 2. Further, the informational barriers also mean that rents accrue, because as soon as a suitable match is found, both sides of the market can save themselves further search costs. To the extent that these rents are shared between employers and employees, this means that the wage rate cannot fall to market-clearing levels.

Although heterogeneity with respect to the skills that are required by the firm and offered by the worker is one of the most important aspects by which workers and firms differ, this aspect will be left until Section 3.3.2. Instead, in Section 3.3.1, workers are assumed to be homogeneous and heterogeneities and informational deficits exist instead regarding the location of jobs and the timing of job creation and job search.

3.3.1 Homogeneous Workers

As stated above, finding an appropriate job-worker pair involves a costly search process on both sides of the labour market. More precisely, if a firm wants to hire a new worker, it must first create a vacancy and enter a time-consuming search process until it finds a suitable employee. In order to simplify the analysis, it is assumed that only the unemployed search for a job,

---

[29] There are in fact two distinctive branches with matching theory. The first branch, considered here, analyses worker and job flows. The second branch is concerned with equilibrium wage dispersion. In these models, search frictions are regarded as the time spent by workers to gather information about wage offers. For general overviews of the development of both branches of matching theories see, for example, Mortensen (1986) and Mortensen and Pissarides (1999b).

i.e. there is no on-the-job search.[30] In this case, an unemployed individual will send applications to all firms who have posted a vacancy for which he finds himself qualified.[31] Each position can have at most one worker.

This search or matching process on the part of firms and workers is modelled by a *matching* function. The number of matches taking place per unit time is given as a function of the number of unemployed and number of vacancies denoted by $u\bar{L}$ and $v\bar{L}$ respectively, where u is the unemployment and v the vacancy rate. Thus, it is implicitly assumed that the intensity of search is not important. Even if this is not a realistic assumption and it is easily possible to include variable search intensities, see, for example, Pissarides (2000, Chap. 5), the additional insights gained from doing so are minimal. Therefore, denoting the number of matches per unit time by $\tilde{m}$, then these can be written as

$$\tilde{m}(u,v)\bar{L} = \tilde{m}(u\bar{L}, v\bar{L})$$

This function is assumed to be increasing in both arguments, concave and homogeneous of degree one.[32] This means that the above function simplifies to

$$\tilde{m} = \tilde{m}(u,v) \tag{3.25}$$

i.e. is independent of the size of the labour force.

At any point in time, job vacancies and the unemployed are matched randomly. Therefore, the rate at which vacancies are filled is

$$f = \frac{\tilde{m}}{v} \tag{3.26}$$

Due to the assumption of a linearly homogeneous matching function, this rate is a function only of the ratio of the vacancy to unemployment rate. For this reason, it is convenient to introduce a labour-market tightness variable, denoted by $\theta \equiv v/u$. With this definition and the properties of the matching function, $f'(\theta) < 0$, i.e. due to congestion externalities, the more firms that are looking to fill vacancies at any moment, the harder will it be for an individual

---

[30] See, for example, Pissarides (2000, Chap. 4) for a model which takes on-the-job search into account. However, as the results of that model do not differ substantially from those without on-the-job search, this extension will not be pursued here.

[31] See Taylor (1995), Coles and Muthoo (1998), Coles and Smith (1998), and Coles (1999) for models where it is assumed that the existing stock of unemployed can only match with the inflow of vacancies and the existing stock of vacancies can only match with newly unemployed individuals.

[32] See Petrongolo and Pissarides (2001) for a survey of the theoretical and empirical literature on matching functions. In this survey, many articles are cited confirming that a matching function which is homogeneous of degree one is consistent with the majority of the empirical evidence.

firm to find a suitable applicant. Hence, θ can be interpreted as an indicator of labour-market tightness. Further, with the probability that a vacant job is matched with a job-seeker during a small time interval dt given by $f(\theta)\,dt$, the mean duration needed to fill a vacancy is $1/f(\theta)$.

The job-finding rate a at which an unemployed individual finds a suitable position is derived analogously to the vacancy-filling rate and is given by[33]

$$a = \frac{\tilde{m}}{u} = f(\theta)\theta \qquad (3.27)$$

so that the mean unemployment duration is $1/(f(\theta)\theta)$. By the properties of the matching function (3.25), $f(\theta)$ has an elasticity in the interval $(-1,0)$. Therefore, due to search externalities, the job-finding rate is an increasing function of labour-market tightness θ as the more vacancies there are relative to job-seekers, the higher will be the total search activities by firms and thus the higher is the probability of finding a suitable job.

The probability that an employed worker is fired due to a job-specific shock is given by the distribution function $\Psi(t)$ which defines the probability that a worker who starts working at t_0, is still employed at a later time t. Formally, this probability is

$$\Psi(t) = e^{-b(t-t_0)}, \qquad b > 0 \qquad (3.28)$$

where b is the job-breakup rate explained below in more detail. From equation (3.28) it can be seen that the conditional probability of someone currently employed still working in the same job after a time interval τ is $\Psi(t+\tau)/\Psi(t) = e^{-b\tau}$ and is thus independent of t. This is an important result as it implies that the probability of a worker who is employed at time t still being employed after the arbitrary time interval τ, is independent of how long the worker has already been employed. Therefore, there is no time dependency and the transition probabilities from one state to another are the same in each period. This kind of "memory less" process is known as a Poisson process. Therefore, the job-breakup rate b is effectively the rate at which jobs are terminated per unit time or alternatively, the conditional probability that a job is terminated at any moment is given by the time constant hazard rate b.[34]

In a steady-state equilibrium, at any moment in time, the number of workers who become unemployed must be equal to the number of job-seekers who find a new job. From the above and without labour force growth, the mean

---

[33] See Blanchard and Diamond (1994), for a model where the job-finding rate negatively depends on unemployment duration.

[34] Thus, it is implicitly assumed that workers can be instantaneously dismissed. For models which analyse the effects that firing restrictions have in more detail see, for example, Saint-Paul (1996a), Garibaldi (1998), Mortensen and Pissarides (1999a) or Cahuc and Postel-Vinay (2002).

number of workers who enter unemployment during a small time interval dt is $b(1-u)\bar{L}\,dt$, and the mean number who leave unemployment in this time interval is $\bar{m}\bar{L}\,dt$ which can be expressed as $f(\theta)\theta u\bar{L}\,dt$. As the labour market is assumed to be large enough so that deviations from the mean can be ignored, the steady-state unemployment rate is derived as

$$u = \frac{b}{b + f(\theta)\theta} \tag{3.29}$$

Before a position can be created, the firm must first decide on the production technology it will employ. Once this decision has been made, it is irreversible and the firm creates a vacancy and actively engages in searching for a worker who meets the job requirements. The job is then created as soon as the job and a worker are matched and have agreed upon the real wage w (as discussed below), and lasts with constant productivity y until it is terminated by the job-specific shock mentioned above.[35]

The number of positions created in the economy is endogenous and determined by profit maximisation. Although it is assumed that a vacancy can be created or terminated at no cost, actively searching for and hiring a worker involves a fixed cost of cy per unit time.[36] Firms are assumed to be risk-neutral and are thus interested in the present-discounted value of expected profits. Denoting this value by W^V for a vacant position and that of a filled position by W^F, and recalling that vacancies are occupied at the rate $f(\theta)$, means that in a continuous time setting, and using dynamic programming, the above value can be calculated to give the Bellman equation for the value of a vacancy as[37]

$$rW^V = f(\theta)(W^F - W^V) - cy \tag{3.30}$$

From this equation it can be seen that a job can be interpreted as an asset that is owned by the firm. Assuming a perfectly competitive capital market, this asset must be valued so that its capital costs are exactly equal to the rate of return of a risk-free investment valued at W^V, where r denotes the interest rate. This asset has fixed costs of cy but with probability $f(\theta)$ it yields a net return of $W^F - W^V$.

---

[35] See Mortensen and Pissarides (1994) for a model where productivity has a global and an idiosyncratic component or Pissarides (1985) for a model with stochastic job matchings where the productivity of a worker is a random draw from a common, cumulative density function.

[36] The assumption that vacancy costs are proportional to the productivity of the worker is not critical if, as here, there is no labour force productivity growth. However, with such growth, this assumption guarantees the existence of a steady state as it ensures that the costs of the firm rise with productivity levels – see Chapter 5.2.3.

[37] These equations are named after the principle of dynamic optimality developed by Bellman (1957). See Appendix A.1 for the derivation in the context of the matching model.

In a steady-state equilibrium with free market entry, all profit opportunities from new jobs are exploited. Therefore, $W^V = 0$ must hold. Using equation (3.30), this implies that the steady-state value of a filled position is

$$W^F = \frac{cy}{f(\theta)} \tag{3.31}$$

i.e. the value of filling a position is equal to the expected search costs. Hence, a tighter labour market leading to a longer and therefore more costly search activity, implies that the value of the newly opened vacancy must rise accordingly so that the expected profit from a new job is always equal to the expected cost of finding a suitable worker. If the job is occupied, the corresponding Bellman equation is given by

$$rW^F = y - w - bW^F \tag{3.32}$$

Again interpreting a filled job as an asset, then this asset yields a net return of $y - w$, but faces the probability b of an adverse shock which leads to a "capital" loss of W^F.

Combining equations (3.31) and (3.32) yields

$$y - w = \frac{r+b}{f(\theta)}cy \tag{3.33}$$

which can be interpreted as the job creation condition or as the demand for labour. The marginal product of labour is given by y and $(r+b)cy/f(\theta)$ is the expected capitalised value of the firm's hiring costs. In a standard neoclassical model of the labour market, firms would have no hiring costs and labour's marginal product is exactly equal to its marginal cost given by the wage rate w. Thus, in contrast to the perfectly competitive labour market and both the union- and efficiency-wage model, the wage in the matching model is less than labour's marginal productivity.

As in this version of the matching model all jobs have the same productivity and all workers search with equal intensity, the job-acceptance decision is trivial. However, the wage rate that a worker receives has a decisive influence on the value of a vacancy to a firm and thus on how many jobs will be created in equilibrium. In order to determine this equilibrium wage, the supply side of the labour market needs to be analysed in more detail.

In a similar fashion, as a vacancy yields a return to the firm, workers earn a return to their human capital. How high this return is, depends on whether they are employed or not. The present-discounted value of lifetime utility for someone employed and earning a wage w is

$$r\tilde{V}^E = w - b(\tilde{V}^E - \tilde{V}^U) \tag{3.34}$$

Intuitively, equation (3.34) states that a worker owns an "asset" which is worth $\tilde{V}^E$ and which pays a dividend of w when he is employed, but faces the probability b of incurring a capital loss of $\tilde{V}^E - \tilde{V}^U$. The value of this asset states how much his human capital is "worth" in monetary terms. Therefore, assuming risk-neutral individuals, this asset must yield at least the same return as a risk-free asset on the capital which pays the interest rate r.

By a similar, argument, the present-discounted value of lifetime utility for an unemployed individual is

$$r\tilde{V}^U = z + f(\theta)\theta(\tilde{V}^E - \tilde{V}^U) \tag{3.35}$$

where z denotes an imputed real return, measured in the same units as real wages, which an unemployed worker obtains. This return may be interpreted as unemployment benefits, income from irregular jobs in a different sector of the economy, or as a fictitious return from unpaid activities such as home production. The value of $r\tilde{V}^U$ can be interpreted as an individuals normal or permanent income, as it is the maximum amount that the unemployed worker can spend without running down his (human) capital. Further, z represents the minimum or reservation wage needed for a worker to accept a job offer.

If a potential firm-worker pair decides not to match, then both will have to go through an expensive search process before forming a new match with a different partner. Therefore, a current match yields an economic rent which is equal to the sum of the expected search costs of the firm and the worker (including forgone wages and profits). The wage paid to the worker determines how this rent is divided up between firms and workers. This wage must be at least as high as the worker's reservation wage, and reaches a maximum at a level at which the firm's surplus from the match reaches zero. The actual wage rate is derived using the Nash bargaining concept as described in Section 3.1.1, only that now bargaining takes place between individual firms and individual workers.[38] On the part of a worker, the surplus that accrues from matching with a firm is the difference in present-discounted value in income when employed compared to being unemployed. For a firm, the surplus is the difference in present-discounted profit streams between a vacant and an occupied position. Therefore, the resulting wage rate is the one that maximises the Nash product derived as[39]

$$\max_{w} \Omega = (\tilde{V}^E - \tilde{V}^U)^{\tilde{\beta}}(W^F - W^V)^{1-\tilde{\beta}} \tag{3.36}$$

where $\tilde{\beta}$ is worker-bargaining power and $\tilde{V}^U$ and W^V can be interpreted as the fallback positions of the worker and firm respectively as was the case when firms and unions bargain over wages.

---

[38] Wage bargaining was first introduced into search theory by Diamond (1982). See Burdett and Mortensen (1998) and Moen (1997) for alternative assumptions about wage setting.

[39] As this Nash product is identical for all firm-worker pairs, specific firm or worker indices are omitted.

Maximisation of the Nash product with respect to the wage rate yields

$$\tilde{V}^E - \tilde{V}^U = \tilde{\beta}(W^F - W^V + \tilde{V}^E - \tilde{V}^U) \tag{3.37}$$

i.e. $\tilde{\beta}$ can also be interpreted as labour's share of the total surplus.

Recalling that in a steady state $W^V = 0$ holds, and inserting equations (3.32) and (3.34) into (3.37) yields

$$w = \tilde{\beta}y + (1 - \tilde{\beta})r\tilde{V}^U \tag{3.38}$$

Combining the steady-state condition (3.31) with equation (3.37) and inserting this into the Bellman equation for the present-discounted value of utility of an unemployed (3.35), leads to expected returns to searching of

$$r\tilde{V}^U = z + \frac{\tilde{\beta}}{1 - \tilde{\beta}}cy\theta \tag{3.39}$$

Inserting this equation into the wage equation (3.38) leads to an equilibrium wage given by

$$w = (1 - \tilde{\beta})z + \tilde{\beta}y(1 + c\theta) \tag{3.40}$$

As can be seen from this wage equation, not only is the wage a positive function of the worker-bargaining power $\tilde{\beta}$ as to be intuitively expected, but also of labour-market tightness θ. This is because a higher labour-market tightness means that it is easier (i.e. requires less search activity) for workers to find new jobs than it is for firms to find new workers. This has the effect of increasing the effective bargaining power that a worker possesses.

The wage rate as given by equation (3.40) together with the flow equilibrium (3.29) and the job-creation condition (3.33), together define the steady-state equilibrium. In order to understand the comparative static results more easily, this equilibrium can be depicted using two figures. By the properties of the matching function, equation (3.33) defines a negative relationship between the wage rate and labour-market tightness. This is depicted as the job-creation curve in Fig. 3.5. The wage curve in Fig. 3.5 is defined by equation (3.40) and is upward sloping for the reasons stated above. The two equations define the unique wage-labour-market tightness equilibrium.

In order to determine this equilibrium in the vacancy-unemployment rate space, equations (3.33) and (3.40) are combined to yield

$$(1 - \tilde{\beta})(y - z) - \frac{r + b + \tilde{\beta}f(\theta)\theta}{f(\theta)}cy = 0 \tag{3.41}$$

This equation is drawn in Fig. 3.6 as the job-creation condition and is a line through the origin with slope θ. Equation (3.29) describes the Beveridge curve

Fig. 3.5: Equilibrium Wages and Labour-Market Tightness

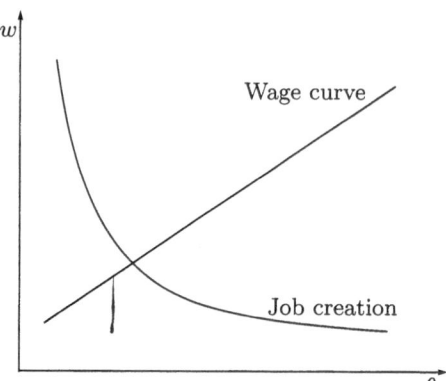

and by the properties of the matching function is convex to the origin. As can clearly be seen from this figure, the higher the number of vacancies, the lower is the unemployment rate as the unemployed will require a shorter time interval in order to find a job. Due to the externalities in the labour market, the

Fig. 3.6: Equilibrium Vacancy and Unemployment Rate

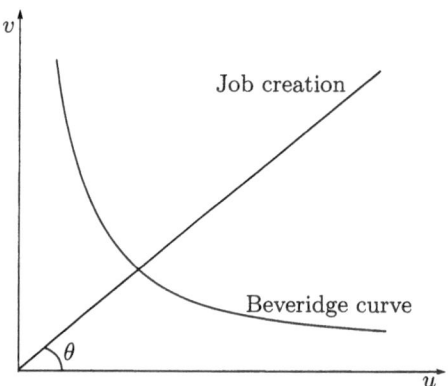

marginal decrease in the unemployment rate is decreasing in the vacancy rate (and vice-versa). Seeing as with very high vacancy rates there are only few unemployed left who are searching for a job, the chances of finding a suitable worker will not increase much by opening up a further vacancy. However, if there is a very high unemployment rate, then one additional vacancy will find a large number of unemployed applying for the position, so that this position will be occupied relatively quickly and the unemployment rate will decrease accordingly.

Turning to the comparative static results, it can be seen that increases in productivity y will shift the job-creation curve in Fig. 3.5 towards the right and the wage curve upwards. It is shown in Appendix A.2 that this productivity increase will unambiguously lead to a higher labour-market tightness. In terms of Fig. 3.6, this increase rotates the job-creation line anti-clockwise. As can be seen from the wage equation (3.40), wages will increase by the less than the productivity level. Therefore, with this higher productivity level, firms' expected profits from opening up vacancies increase, so that in the new equilibrium, the job-creation rate is higher. As can also be seen from Fig. 3.6 and equation (3.29), this higher labour-market tightness will lead to a reduction in the unemployment rate.

If workers are able to increase their bargaining power $\tilde{\beta}$, then the job-creation condition (3.33) and hence the job-creation curve in Fig. 3.5 remain unaltered. However, as can be seen from the wage equation (3.40), the wage curve rotates anti-clockwise so that the new equilibrium is characterised by lower labour-market tightness. Therefore, the job-creation condition shown in the $u - v$-space in Fig. 3.6 will rotate clockwise. Further, the lower value of θ also means that the Beveridge curve shifts outwards. Hence, even if the effect on the vacancy rate is ambiguous, the unemployment rate will increase.

Higher values of the interest rate r lower the present-discounted values of expected profits. As can be seen from the job-creation condition (3.41), this reduces the number of vacancies and thus leads to a fall in labour-market tightness θ. In terms of Fig. 3.6, this lower labour-market tightness shifts the job-creation line clockwise so that the unemployment rate is now unambiguously higher.

From equation (3.29) which defines the Beveridge curve and the job-creation condition (3.41), it can be seen that a higher job-destruction rate b leads to a clockwise rotation of job-creation line and to an outward shift of the Beveridge curve so that the unemployment rate will be higher in the new equilibrium. That unemployment increases with a higher job-destruction rate is to be intuitively expected. If more jobs are destroyed in a certain time interval so that the inflow into unemployment increases, flow equilibrium can only be reached if the outflow from unemployment rises accordingly. With a lower labour-market tightness implying a longer unemployment duration, this increase in outflow can only be reached if the number of unemployed searching for a job increases. Further, the increase in the job-destruction rate and the resulting fall in labour-market tightness will also lead to a reduction in the wage rate as the duration of a job and therefore the expected surplus from the match decrease.

From equation (3.29) it can also be seen that an increase in hiring costs c have no direct influence on the Beveridge curve. However, higher vacancy costs negatively affect the job-creation condition (3.41) as firms will reduce the number of vacancies as the expected revenue from a vacancy must rise in order to compensate for the increasing costs. In terms of Fig. 3.6, this rotates the job-creation line clockwise and the unemployment rate rises.

Finally, an increase in unemployment benefits z will lead the wage curve in
Fig. 3.5 to shift upwards, whereas the job-creation curve remains unchanged.
Thus, the wage will rise but labour-market tightness falls. In Fig. 3.6, the
job-creation curve rotates clockwise whilst the Beveridge curve remains unaffected. Hence, the unemployment rate increases whereas the vacancy rate
falls.

The position of the Beveridge curve in Fig. 3.6 also depends on the matching
technology as given by (3.25). As stated at the beginning, both sides of the labour market have imperfect information about job-offers and job-seekers. The
efficiency with which this information is transmitted is given by the matching technology. Therefore, if workers or jobs become "more heterogeneous",
i.e. the degree to which each party has imperfect information increases, then
for a given matching technology, search activities must also increase so that
each vacancy rate is now associated with a higher unemployment rate. Thus,
the larger is the divergence between the qualifications that firms require and
job-seekers possess, the higher is the degree of mismatch. Since a higher degree of mismatch also reduces the rate $f(\theta)\theta$ at which the unemployed find
jobs, labour-market tightness declines, i.e. the job-creation line in Fig. 3.6
rotates clockwise and, as can be seen from equation (3.29), the Beveridge
curve shifts outwards. Thus, the Beveridge curve is also an indicator of the
degree of "mismatch" on the labour market. It is such a higher mismatch,
particularly amongst the low-skilled, that is often stated as being the cause
of the labour-market problems for this type of labour. A reason that is often stated for this increase in mismatch is the nature of recent technological
change. Although leaving a discussion of why technological change has been
biased towards high-skilled workers until Chapter 6.3.3, it is obvious that it is
important to analyse matching theory in the context of heterogeneous labour.

3.3.2 Heterogeneous Workers

The above analysis can easily be modified to incorporate labour which is
heterogeneous with respect to skills. As throughout, the distribution of skills
across workers is assumed to be exogenous so that the fraction of the workforce
that is high-skilled is given.[40] Assuming that vacancies are announced so that
they are equally visible to both the low- and high-skilled but that all vacancies
are skill specific, means that the matching technology will be the same for both
types of labour and only the arguments of the matching function differ

$$\tilde{m} = \tilde{m}(u_s, v_s), \qquad s \in \{L, H\} \tag{3.42}$$

---

[40] See McKenna (1996) for a matching model in which the role of different education
is modelled explicitly and Acemoglu (1999) for a model where education and skills
are not perfectly correlated.

As will be shown below, the high-skilled will only search and apply for high-skilled positions and as a low-skilled worker will never be accepted for such a job, the two types of labour are active in separate labour markets.[41] Therefore, the vacancy filling and job-finding rates will be of the same functional form as given by equations (3.26) and (3.27), but also be skill-specific and are now given by $f(\theta_s)$ and $f(\theta_s)\theta_s$ respectively. Further, the flow equilibrium derived in equation (3.29) must also hold for each sector. In addition, it is assumed that each type of labour has a unique and independent probability b_s of receiving a negative shock leading to the termination of a job. Therefore, the Beveridge curve is specific to each sector and given by

$$u_s = \frac{b_s}{b_s + f(\theta_s)\theta_s} \tag{3.43}$$

Although it is not a priori clear whether the job-breakup rate b_s should be higher or lower for the low-skilled, empirical studies find that it decreases with skills, i.e. $b_L > b_H$ (see Gautier et al. 1999), so this is the case that will be assumed here.

The derivation of the job-creation condition is analogous to above with the exception that equations (3.30) – (3.41) now need to be augmented by the skill index s. Thus, if both low- and high-skilled unemployed individuals place the same "value" z on leisure, and if the productivity-adjusted hiring costs c are the same for both types of vacancies, the job-creation condition for each skill-type now becomes

$$(1 - \tilde{\beta}_s)(y_s - z) - \frac{r + b_s + \tilde{\beta}_s f(\theta_s)\theta_s}{f(\theta_s)} cy_s = 0 \tag{3.44}$$

and the wage curves are given by

$$w_s = (1 - \tilde{\beta}_s)z + \beta_s y_s(1 + c\theta_s) \tag{3.45}$$

Productivities within each skill-group are assumed to be homogeneous with the high-skilled having a higher productivity than the low-skilled, $y_H > y_L$.[42] As shown in Appendix A.2, labour-market tightness increases with productivity levels, so that the market for high-skilled workers is characterised by a higher ratio of vacancies to unemployed, $\theta_H > \theta_L$. In addition, their lower job-destruction rate b_H works in the same direction and leads to a further

---

[41] The assumption that a low-skilled worker will never be accepted for a high-skilled vacancy can be justified on the grounds that, as the production technology decision occurs before the vacancy is created, the two technologies differ so much, that only the high-skilled can operate the machines in operation in this sector.

[42] See Kohns (2000) for an extension of the Mortensen and Pissarides (1994) model with heterogeneous labour where productivity has a global and an idiosyncratic component.

increase in high-skilled labour-market tightness. Hence, as can be seen from the Beveridge curve (3.43) and the fact that the job-finding rate is an increasing function of labour-market tightness, the high-skilled unambiguously have a lower unemployment and higher job-finding rate than the low-skilled. Further, as can be seen from the wage equation (3.45), the higher labour-market tightness and productivity of the high-skilled also imply that their wages are higher. Although it is possible that the low-skilled workers have a higher bargaining power β_L than the high-skilled – the low-skilled are more likely to be represented by a union and not bargain on an individual level – there are also arguments for assuming that the high-skilled have a higher bargaining power – their higher productivity and resulting higher labour-market tightness increases their effective bargaining power. Which effect dominates cannot be determined analytically and is an empirical question. However, it is assumed here that even if the low-skilled have a higher-bargaining power, this effect is not strong enough to push low-skilled wages above the high-skilled level. Therefore, with wages higher and unemployment duration lower for the high- than the low-skilled, the high-skilled will always only search for high-skilled jobs as the returns to searching are highest in this case.[43]

As can be seen from the equilibrium equations (3.43), (3.44) and (3.45), the equilibrium for both types of labour is independent of the size of either workforce. This result is due to the fact that the matching function is assumed to be of a constant returns to scale type, so that only labour-market flows and the unemployment and vacancy *rates* are decisive. However, even if the average unemployment rate is independent of the size of the workforce, the composition does play a crucial role. Thus, c.p., an increase in the fraction of workers who are highly skilled will lower the average unemployment rate as now a larger fraction of the population is of the skill type with the lower unemployment rate. However, this result depends on the assumption that there are no strategic connections between firms posting vacancies on the one hand and the two types of labour on the other, so that there are in fact two separate labour markets. This assumption is dropped in, for example, Acemoglu (1999) and Saint-Paul (1996a) where firms choose which type of vacancy to post depending on the number of highly skilled in the economy. In this case, an increase in the fraction of highly skilled workers will also lead to fewer vacancies being posted for the low-skilled, so that their unemployment rate will increase so that the precise effects on the average unemployment rate are now ambiguous.[44] These links between the labour market for the low- and

---

[43] See, for example, Albrecht and Vroman (2002) or Gautier (2002) for models in which high-skilled workers also take-up low-skilled jobs. In this case it is possible that low-skilled unemployment rises as high-skilled workers crowd out the low-skilled. See, for example, Gautier et al. (2002) for empirical evidence of such crowding out.

[44] In Acemoglu (1999) it is even possible that the low-skilled market collapses completely so that all low-skilled are unemployed.

high-skilled respectively are explored in more detail in Chapter 6.3.3 where the size of the workforce of each type of labour influences the direction of technological research and thus the job-creation condition.

3.3.3 Discussion

Chapter 3.3 has introduced the class of search or "matching" models in a basic form with homogeneous workers and for the case when there are low- and high-skilled workers. In contrast to the unrealistic assumption in rudimentary neoclassical labour-market models where both firms and job-seekers are perfectly informed about all employment prospects, matching models assume that, due to imperfect information about the exact job locations, requirements or qualifications needed, market frictions occur. Therefore, the unemployed must search through all job advertisements and firms must interview applicants before a suitable firm-worker pair can be found or matched. Once a potential match has been formed, the potential worker and firm form a bilateral monopoly and bargain over the wage rate which determines how the surplus accruing from the match is distributed. Therefore, with the worker having some degree of bargaining power, similar to the models of union wage bargaining discussed in Chapter 3.1, the wage rate will never reach its market-clearing level. However, in contrast to the union wage-bargaining models, even in the extreme situation that a worker has no bargaining power, due to the fact that with imperfect information firms incur search and hiring costs, the wage rate cannot fall enough to clear the labour market. Thus, labour market frictions mean that even perfectly flexible economies with no wage floors will still be characterised by a certain degree of unemployment.

Clearly, the efficiency of the matching function plays a crucial role in the model. The faster information from one side of the labour market reaches the other, the lower are the search frictions and the steady-state unemployment rate. Thus, even if it need not automatically be the government that sets up employment agencies, from an efficiency point of view, this could just as easily be undertaken by private job centres, the government clearly has an intrinsic interest that such agencies exist and function well. To this extent, at the extreme, it can be argued that there is no need for many different job centres and private agencies each having only an overview over a limited number of jobs and workers, but especially in the days of the internet, it is more efficient if there is only one database containing all currently available positions and job-seekers. Further, as shown above, the unemployment rate declines with increasing job productivity. Thus, although such policies may only be effective in the long-run, any investment the government undertakes in its education system leading to a more highly qualified workforce, will reduce future unemployment.

As shown above, the imputed real income of the unemployed influences how high the equilibrium wage and therefore unemployment will be. Thus, interpreting this income as unemployment benefits, means that one policy recommendation resulting from the matching theory is to reduce unemployment benefits. In fact, as shown for example in Hansen and İmrohoroğlu (1992), this policy recommendation gains further weight if the disincentive effects of benefits on search intensity and job acceptance are explicitly modelled. However, there are by now numerous models which concentrate on the positive insurance and, as with higher benefits the unemployed can search longer in order to find a better match, the productivity-enhancing effects of benefits. Thus, even if one were to ignore the social consequences of such a policy measure, the effects on unemployment of reducing benefit payments is by no means as clear cut as it may seem at first.[45]

Undoubtedly, the matching models in this chapter can explain many empirical phenomena such as the coexistence of vacant positions and unemployment or the fact that search frictions mean that even labour markets with no wage floor will not clear. However, as the rate at which jobs are destroyed is exogenous, a key element in explaining how high the unemployment rate is (on an aggregate as well as skill-specific level), is left unexplained for the moment. This assumption is dropped in Chapter 6.3.3

3.4 Summary and Conclusions

This chapter has presented three different types of labour-market imperfections which all have in common that they can explain why the wage level does not fall to market-clearing levels even in the presence of unemployment. In Chapter 3.1, it was unions representing the labour force and hence acting as a labour-supply monopolist who had enough bargaining power to prevent wages from falling. Chapter 3.2 showed that employers themselves have an intrinsic interest at not letting the wage fall to market-clearing levels because higher wages also encourage workers to exert more effort and hence to produce at higher productivity levels. Finally, the matching theory of unemployment introduced the notion that labour markets are characterised by search frictions, i.e. both employers and employees must undergo a costly search process before finding each other. Once a worker and firm find each other, they can save themselves further search costs by forming a (temporary) match. It is

---

[45] For a more detailed discussion and theoretical model see Fredriksson and Holmlund (2001) and the papers cited therein. Gruber (2001) provides empirical evidence how the insurance aspect varies with the wealth of the unemployed and shows that it can actually be optimal to increase benefits with unemployment duration.

these rents due to the presence of frictions so that vacancies cannot be instantly filled which determine the wage rate and prevent the labour market from clearing.

The models in this chapter are designed so as to present the foundations for more complex and realistic models in the subsequent chapters. For this reason, many important aspects such as economic growth and technological change have not been taken into account so far. This will be done in Chapter 5 where economic growth through exogenous technological change is analysed and in Chapter 6 where the cause of long-run growth is endogenised and the effects of skilled-biased technological change are studied.

However, before growth is integrated into the above models, two further shortcomings of all the models discussed so far will be addressed. Firstly, that the goods market is characterised by perfect competition. Not only is this assumption unrealistic, but as a firm's demand for labour is derived from the demand for the goods or services which it produces, it also immediately becomes obvious that the two markets are inherently connected to one another so that the degree of product-market competition will have a direct influence on the unemployment rate. Secondly, the results derived so far have all been within a partial equilibrium context. However, this means that all effects which indirectly influence the outcome of the labour market, cannot be analysed.

4

Market Structure and Unemployment

Two features characterise the models discussed in the previous chapter: first, they all assume a perfectly competitive product market, and second, without a full general-equilibrium analysis, both the price of as well as the demand for the good or service that is being produced is exogenously given. Although these assumptions can be justified if the aim of the model is to focus only on the labour-market distortions, they also mean that potentially important aspects on other markets which indirectly affect unemployment are either neglected or cannot be properly analysed. That the integration of imperfectly competitive product and labour markets into a dynamic general-equilibrium model is of importance to understand the causes of unemployment was first emphasised by Phelps (1994, 1995).[1] For example, the price a firm charges for its good or service and the budget restraint that consumers face both play a decisive role in determining the demand for this good and hence the demand for labour. It is for this reason that the interactions between the product and labour markets are the focus of this chapter.

Also of importance in the models in this chapter is the fact that empirical studies find that there are significant wage differences between different sectors of the economy even if individual and workplace aspects such as qualifications, age, race, sex and working conditions are taken into account.[2] In other words, homogeneous workers receive different wages depending on which sector they are employed in. As these inter-industry differences have been persistently measured for many years, they are obviously not eroded away by competitive forces. This suggests that there are systematic differences between different sectors of the economy. For this reason, the models in this chapter do not treat the labour market as one unified sector, but adopt the so-called "segmented"

---

[1] See Stadler (1996) for an evaluation of this class of model.

[2] See Dickens and Katz 1987, Krueger and Summers 1988, Katz and Summers 1989, OECD 1994b and Abowd and Kramarz 2000 who also provide a detailed discussion of the measurement problems in measuring these differences.

or "dual" labour-market approach and differentiate between two sectors of the labour market.[3]

This approach was first developed by Harris and Todaro (1970) and subsequently modified in seminal articles by Calvo (1978), McDonald and Solow (1985) and Bulow and Summers (1986). In this class of models, the labour market is dichotomised into a primary sector where, by definition, wages are higher, there is a higher degree of job security, the turnover rates are lower and a secondary sector, where the opposite holds. Although in reality, of course, the labour market can be segregated into a whole continuum of sectors, the simplifying assumption of only two sectors has proven to be a good approximation with a large body of empirical evidence in support of the theory (see, for example Haisken-DeNew and Schmidt 1999 and Dickens and Lang 1993, as well as the survey in Saint-Paul 1996b). These difference between the two sectors imply that wages are attached more to jobs than to workers, with all "good" jobs in one sector.

In order to highlight the differences between the sectors, it is assumed that the secondary sector produces a homogeneous traditional good or service and is comprised of menial jobs for which the wage is determined by market clearing.[4] The primary sector produces a composite manufacturing good and is characterised by imperfect product-market competition so that rents will accrue here. As a result, if labour is able to obtain some of this economic surplus, then wages in this sector will be above their market-clearing level and an equilibrium non-competitive wage differential exists between the two sectors. These higher wages in the primary sector will induce agents currently not employed in this primary sector to be prepared to spend time (in which they are unemployed) applying for these jobs. Therefore, increased competition on the product market which leads to lower rents and thus also to a lower wage differential, should also lead to lower unemployment as the rewards from searching for such a job decrease. In addition, the fall in prices associated with more intense competition also increases product demand and hence increases the employment rate in the high-wage sector. This is exactly the idea stated in the OECD Jobs Study (OECD 1994b, p. 23) and presented more formally below. As shown in Chapter 2, recent work by the OECD (OECD 2002, Chap. 5) also finds convincing empirical evidence in support of this hypothesis.

For the reasons outlined above, rents accrue in the primary sector. In Chapter 4.2.1, these rents lead workers to unite to form a union in order to extract a larger share than an individual worker bargaining on his own could. Chapter 4.2.2 analyses the case when firms are willing to share these rents with the workers due to efficiency-wage considerations as only in this case are employ-

---

[3] See Wapler (1999) for a survey.

[4] Even in Europe, market clearing in the secondary labour market is not contradicted by empirical evidence (see Dolado et al. 1996). However, see Jones (1987a) for a model with a binding minimum wage level in the secondary sector.

ees prepared to provide effort. Finally, Chapter 4.2.3 analyses the case when the labour market is characterised by matching frictions. However, before analysing the labour market in more detail, the optimal household and producer decisions which hold irrespective of the imperfections present in the labour market, will first be analysed in more detail.

4.1 Household and Producer Decisions

The economy consists of $\bar{L}$ homogeneous and risk-neutral workers who are allocated across the sectors as follows

$$\bar{L} = \bar{L}_T + \bar{L}_M \tag{4.1}$$

$$\bar{L}_T = L_T, \qquad \bar{L}_M = L_M + U_M \tag{4.2}$$

where $\bar{L}_T$ denotes the size of the workforce in the secondary or traditional sector. As the labour market in this sector is assumed to be perfectly competitive and hence always clears, the employed labour force L_T, and the total number of workers in this sector $\bar{L}_T$, coincide. The total number of individuals in the primary or manufacturing sector is given by $\bar{L}_M$ and consists of L_M who are employed and U_M who are unemployed.[5] These unemployed choose to wait for a high-paying job, i.e. they decide not to take up a job in the low-wage sector. This is in accordance with the empirical evidence, that although unemployment is a bad signal, being in a low-wage job may well be an even worse signal.[6] Further, if workers differ with respect to their preferences for primary-sector work, then those with higher preferences for these jobs will be more prepared to queue for this type of work and hence, will require a lower wage to achieve a given utility level.

4.1.1 Households

Each household is treated as an infinitely-lived dynasty. All dynasties are assumed to have the same discount rate and identical preferences so that intertemporal utility U is given by

$$U(C) = \int_0^\infty e^{-\rho t} \frac{C_t^{1-\gamma} - 1}{1 - \gamma} \, dt \tag{4.3}$$

---

[5] The terms secondary and traditional sector on the one hand, and primary and manufacturing sector on the other, will be used synonymously throughout.

[6] See Laing (1993), Ma and Weiss (1993) and McCormick (1990), for theoretical models of signalling and job search with theoretical and empirical support of this claim.

where $\gamma > 0$ is the elasticity of marginal utility and C_t is total consumption at time t.[7] Therefore, $1/\gamma$ denotes the intertemporal elasticity of substitution and is a measure of the extent to which households are willing to have different consumption levels over time.

Preferences for the goods produced in the traditional and manufacturing sector respectively are given by the Cobb–Douglas function

$$C_t = M_t^\zeta T_t^{1-\zeta} \tag{4.4}$$

where M_t and T_t are the consumed quantities of the manufacturing and traditional good respectively, and ζ is the expenditure share spent on manufacturing goods. M represents a composite good comprised of n varieties. As will be shown below, each variant is produced by a single firm, so that the number of varieties and the number of firms operating in the manufacturing sector are equal.

Households' preferences for the manufacturing goods are given by a CES utility function

$$M = \left[\sum_{i=1}^n m_i^\kappa\right]^{\frac{1}{\kappa}}, \qquad 0 < \kappa < 1 \tag{4.5}$$

with m_i as the consumed quantity of the manufacturing good of brand i.[8] Here, κ is a measure of the preference for variety or also of the homogeneity of the goods. Assuming symmetrical firms so that consumers have equal preferences for all brands means that as κ approaches one, the differentiated goods become almost perfect substitutes for one another and total utility gained from consuming the composite good is simply the sum of the consumed amounts of each variant. As, however, κ decreases towards zero, the goods become more varied so that households gain in utility simply due to the fact that they now have more differentiated products to choose from. It is for this reason that households are said to have a "love of variety". Thus, a fall in κ means that the households can attain a given utility at a reduced consumption level. In addition, it can also be seen from the household preferences for the manufacturing goods as given by (4.5), that utility is an increasing function of the number of brands produced. Therefore, households will always choose to consume all available brands and a new firm entering the market always has an incentive to produce a new variety instead of competing with an existing producer in which case demand for this variant will be met by two producers each producing less and charging a lower price than they otherwise could. Hence, a firm's

---

[7] The techniques needed to solve such an intertemporal household maximisation problem go back to the seminal work of Ramsey (1928). For the special case of $\gamma = 1$, the utility function approximates to $U(C) = \int_0^\infty e^{-\rho t} \ln C_t \, dt$.

[8] This type of specification for preferences was developed by Spence (1976) and Dixit and Stiglitz (1977).

profits will always be higher by producing a new variant of the composite good so that each firm has a monopoly on the variant it produces. However, as the individual variants are imperfect substitutes for one another, each firm faces a certain degree of competition from the other firms and is thus limited in the price it can charge. It is for this reason that this kind of imperfection on the goods market is labelled "monopolistic competition".

Consumers face a three stage optimisation problem. In a first step, they must decide how to divide total income between savings and consumption. Formally, households maximise utility as given by (4.3) subject to the intertemporal budget constraint

$$\dot{G} = r_t G_t + I_{w_t} - P_t C_t \tag{4.6}$$

where G denotes household assets, I_w is average wage income, P is the macroeconomic price index defined below in more detail and a dot over a variable denotes the derivative with respect to time t.

Solving this intertemporal optimisation problem results in the Keynes–Ramsey rule for the rate of consumption growth[9]

$$\frac{\dot{C}}{C} = \frac{1}{\gamma}\left(r - \frac{\dot{P}}{P} - \rho\right)$$

As there is no technological progress, the macroeconomic price index P is a constant so that the equation simplifies to

$$\frac{\dot{C}}{C} = \frac{1}{\gamma}(r - \rho) \tag{4.7}$$

Therefore, the relationship between the interest rate r and the rate of time preference ρ determines whether households choose a pattern of consumption that rises or falls over time. A lower inclination to substitute intertemporally, i.e. a higher value of γ implies a smaller responsiveness of the rate of consumption growth to changes in the gap between r and ρ. In the steady state analysed here and without technological growth, steady-state consumption will be constant so that

$$r = \rho \tag{4.8}$$

holds, i.e. the equilibrium is characterised by a constant real (and nominal) interest rate.

In a second stage, in each period (so that the time index t can be omitted without loss of information) consumers optimally allocate their consumption expenditures between manufacturing and traditional goods, that is they choose

---
[9] See Appendix A.3 for the derivation.

M and T so as to maximise the consumption function (4.4) subject to the budget constraint

$$p_M M + p_T T = PC$$

where p_T is the price of the traditional good and p_M the price index of the composite manufacturing good which is derived below.

The result of the second stage optimisation problem is

$$p_M M = \zeta PC \tag{4.9}$$

and

$$p_T T = (1 - \zeta)PC \tag{4.10}$$

as the expenditure shares spent on manufacturing and traditional goods respectively and

$$P = \left(\frac{p_M}{\zeta}\right)^{\zeta} \left(\frac{p_T}{1-\zeta}\right)^{1-\zeta} \tag{4.11}$$

as the macroeconomic price index.[10]

In a third step, consumers decide how to divide their total spending on manufacturing goods amongst the n variants. This demand for variant i is derived in Appendix A.5 as

$$m_i = \frac{p_{m_i}^{-\sigma}}{p_M^{-(\sigma-1)}} \zeta PC, \qquad i = 1, \ldots, n \tag{4.12}$$

where p_{m_i} denotes the price demanded by firm i in the manufacturing sector, $\sigma \equiv 1/(1 - \kappa)$ the elasticity of substitution between any two variants which, with the specification of the subutility function as given by (4.5), is constant for all pairs and independent of the absolute consumption level. Finally, p_M, the price index of the composite manufacturing good is derived as

$$p_M \equiv \left[\sum_1^n p_{m_i}^{1-\sigma}\right]^{\frac{1}{1-\sigma}} \tag{4.13}$$

This price index can be interpreted as the minimum cost of purchasing one unit of the composite good M. Assuming that the number of firms n is "large" means that the effect on the manufacturing-sector price index p_M of a change in the price charged by a single firm p_{m_i} is so small that it can be neglected. In this case, the (absolute) price elasticity of demand for each variety is constant and equal to σ.[11]

---

[10] See Appendix A.4 for the derivation.
[11] As can be seen from the above definition of σ, the elasticity of substitution needs to be larger than one to ensure that marginal revenue is not negative. However,

4.1.2 Firms

Each firm in the traditional sector produces according to an identical technology, with aggregated output given by

$$T = L_T \tag{4.14}$$

This means that labour in this sector has a constant unitary marginal productivity. Seeing as firms in this sector face perfect competition and labour is the only input in production, it must hold that producers in this sector set their price equal to their marginal costs

$$p_T = w_T \tag{4.15}$$

where w_T is the nominal wage rate paid in the traditional sector. In the following, labour productivity and hence the price level in this sector are normalised to one.

Assuming as in the previous chapter that the capital stock is constant and equal to one, means that the production technology of a single firm in the manufacturing sector is specified as

$$m_i = A L_{m_i}^{\alpha}, \quad A > 0, \quad 0 < \alpha < 1 \tag{4.16}$$

where A is a parameter denoting the technological level, α is labour's output elasticity, and L_{m_i} is the amount of labour employed by firm i in the manufacturing sector. Further, firms in this sector also incur fixed costs $\tilde{F}$ in the form of forgone output where, in the following, the value of α and $\tilde{F}$ are chosen so as to guarantee that firms face decreasing average costs. These assumptions together with a free-entry condition ensure that the number of firms is finite.

Firms choose labour input so as to maximise their present-discounted value of future profits given by

$$\max_{L_{m_i}} \int_0^\infty \pi_{m_i} \mathrm{e}^{-rt} \, \mathrm{d}t = \max_{L_{m_i}} \int_0^\infty [p_{m_i}(m_i)m_i - w_{m_i} L_{m_i} - \tilde{F}] \mathrm{e}^{-rt} \, \mathrm{d}t \tag{4.17}$$

where w_{m_i} is the wage paid by firm i in the manufacturing sector. This optimisation problem is identical in every period so that the time index t can be omitted without loss of content. Profit maximisation leads to

$$p_{m_i} = \frac{1}{\kappa} \frac{w_{m_i}}{\partial m_i / \partial L_{m_i}} \tag{4.18}$$

Thus, $1/\kappa$ denotes the markup factor by which prices exceed marginal costs and hence is both a measure of the heterogeneity of goods and also indicates

---

since σ is the substitution elasticity between pairs of varieties of the same product, the expected value is large so that the above assumption is not a severe restriction.

the degree of product-market competitiveness. A higher value of κ implies that firms can only charge a lower markup over their marginal costs and therefore that product-market competition is more intense, with $\kappa = 1$ as the special case of perfect competition.

This completes the description of the household sector and the firm pricing behaviour. In order to determine the wage rate paid in the manufacturing sector, it is now necessary to introduce the labour market where we again differentiate between the three types of wage setting analysed in Chapter 3.

4.2 Wage Setting and General Equilibrium

Chapter 3 showed how different labour-market imperfections prevent the wage rate from falling and hence lead to the existence of equilibrium unemployment. In addition to these imperfections, now due to the assumption of monopolistically competitive firms in the manufacturing sector, the product market is also characterised by imperfect competition so that firms are able to charge a price above their marginal costs. This means that they have the ability to earn rents in this sector. It is the existence of these rents which provide workers with incentives to become members of a union as analysed in Chapter 4.2.1. This union will have additional bargaining power compared to an individual worker when it comes to negotiating the wage rate and hence a greater ability to force the employers to share the economic surplus with the workforce. Chapter 4.2.2 analyses how imperfect competition on the product market alters equilibrium wages and unemployment in the context of an efficiency-wage model. Finally, in Chapter 4.2.3, the interaction of monopolistic competition and matching frictions and how these alter the surplus from forming a match are the subject of attention.

4.2.1 Union Wage Bargaining

In a standard neoclassical model of the labour market, the wage rate across sectors for homogeneous workers must be identical. However, as stated in Chapter 3.1, there is convincing empirical evidence, that a positive wage differential exists between unionised and non-unionised sectors. Whereas in the previous chapter it was the assumption about the production technology that allowed unions to successfully operate, in the dual labour-market model analysed in this chapter, it is the rents due to the imperfectly competitive product market that allows unions to operate and which, as will be shown below, prevents wages from equalising across the sectors.

Calvo (1978) was the first to analyse union wage behaviour within a dual labour market. However, in that model, the goods markets are both assumed to be perfectly competitive. This assumption was modified by Hart (1982) who analysed the interaction of unions and price-setting firms. However, he assumes monopolistic competition as formulated by Chamberlin (1933), which means the number of firms is exogenous and several strong assumptions about the demand functions need to be made in order to obtain a unique equilibrium. More recently, the combination of unions and imperfect product markets has been taken up by Dutt and Sen (1997) and Arnsperger and de la Croix (1990). However, in contrast to here, both of these models only assume a single sector labour market. There are also several recent papers which analyse various affects of unions within a dual labour-market setup. Roberts et al. (2000) analyse a two-stage bargaining process in which unions first determine a national minimum wage and then subsequently the wage in the primary sector. A dynamic innovation-based growth model is developed by Stadler (1999). The original Calvo (1978) model is extended by Dixon et al. (1999), who incorporate a standard menu cost setup. Finally, Burda (1988), analyses how unions affect unemployment duration. However, apart from the model by Stadler (1999), all of these papers have in common that the product market is treated as being perfectly competitive. However, this is, of course, not only an unrealistic assumption, but also overlooks the fact that the labour and product market are uniquely interdependent.[12] These drawbacks are overcome below in which a dual labour-market model with imperfectly competitive goods markets is analysed within different wage-setting schemes.

4.2.1.1 The Model

As shown by equation (3.9) which determines the equilibrium wage which results from bargaining between employers and unions, the elasticity of labour demand with respect to wages is decisive. With imperfect competition on the product market so that a firm can influence its price by the quantity it produces, the relationship between the size of the profit-maximising labour force and the wage rate can be found by totally differentiating the pricing equation (4.18) with respect to labour and wages. Assuming symmetrical firms, making it possible to drop the index i, gives

$$\kappa \frac{\partial p_m}{\partial m} \frac{\partial m}{\partial L_m} \, \mathrm{d}L_m + \frac{w_m \partial^2 m/\partial L_m^2}{(\partial m/\partial L_m)^2} \, \mathrm{d}L_m - \frac{1}{\partial m/\partial L_m} \, \mathrm{d}w_m = 0$$

which by using the production function as given by (4.16) and rearranging yields

---

[12] Stadler (1999) assumes a perfectly competitive consumer goods sector and an imperfectly competitive intermediate goods sector. Whereas in that model, market power arises due to innovative activities, here it stems from monopolistically competitive firms.

$$\frac{\mathrm{d}L_m}{\mathrm{d}w_m}\frac{w_m}{L_m} \equiv \epsilon_{L_m,w_m} = \frac{-1}{1-\alpha\kappa} \tag{4.19}$$

where ϵ_{L_m,w_m} is the elasticity of labour demand with respect to wages in the manufacturing sector.

Equation (4.19) shows that a reduction in product-market competitiveness, i.e. a lower value of κ, reduces (in absolute terms) the elasticity of labour demand. This is because lower pressure from competitors means that firms can demand higher markup prices. Therefore, if wages increase, firms do not have to bear the total burden of these increased costs by dismissing workers, but can instead pass on some of the higher costs to the consumer. The more market power a firm has, the higher is the share of the burden that is passed on to consumers and the lower is the share that the firm itself has to bear, i.e. the lower is the number of workers that are dismissed.

For the reasons specified in Chapter 3.1, unions derive utility from wages and employment levels. In order to be able to derive tractable results, the functional form of the union objective function is specified. This also helps separate union wage effects from the effects on wages due to product-market imperfections. Specifically, as all workers can find a job in the traditional sector at any time, the wage there represents the fallback wage and the unions' aim is to negotiate higher manufacturing-sector wages w_m than those paid in the traditional sector, i.e.

$$\max_{w_m} \tilde{V} = \int_0^\infty [L_{m_t}(w_{m_t} - w_{T_t})]\mathrm{e}^{-\rho t}\,\mathrm{d}t \tag{4.20}$$

where $\tilde{V}$ represents intertemporal union utility and ρ is the union discount rate.[13] As the real wage rate will be constant in a steady state, the time index t can be omitted. Thus, during the wage bargaining process, a necessary condition for unions to maximise their per-period utility V is

$$\frac{\mathrm{d}V}{\mathrm{d}w_m} = \frac{\partial V}{\partial w_m} + \frac{\partial V}{\partial L_m}\frac{\mathrm{d}L_m}{\mathrm{d}w_m} \overset{!}{=} 0 \tag{4.21}$$

There are two opposing effects in (4.21). On the one hand, increases in the wage level directly increase union utility. On the other hand, there is an indirect effect as the size of the workforce will decline with higher wages. As it is assumed that all workers are also union members (or at least that unionised and non-unionised workers receive the same wage), this latter effect will lead to a reduction in union utility. In the optimum these two counteracting effects need to be equalised.

As explained in Chapter 3.1, the most common way of modelling wage negotiations is the (generalised) cooperative Nash bargaining solution. If no

---

[13] As the unions represent the interests of their members, they have the same subjective discount rate ρ.

agreement is reached, firms have a (negative) fallback position $\pi_0 = -\tilde{F}$ due to their fixed costs in form of lost output. For this reason, the Nash product only has variable profits π_V net of the fixed costs, i.e. $\pi_V \equiv \pi + \tilde{F}$ as one of its arguments. For the unions, the fallback position is the wage rate paid in the traditional sector which all workers could receive at any time.[14] Therefore, the Nash product is now given by

$$\Omega = V^\beta (\pi_V)^{1-\beta} \tag{4.22}$$

Maximising this with respect to the wage rate in the manufacturing sector yields

$$\frac{\partial \Omega}{\partial w_m} = \beta V^{\beta-1} \pi_V^{1-\beta} \frac{\partial V}{\partial w_m} + (1-\beta) V^\beta \pi_V^{-\beta} \frac{\partial \pi_V}{\partial w_m} \stackrel{!}{=} 0 \tag{4.23}$$

Using the envelope theorem by which $\frac{\partial \pi_V}{\partial w_m} = -L_m$, means that (4.23) simplifies to

$$\pi_V \frac{\partial V}{\partial w_m} = \frac{1-\beta}{\beta} V L_m \tag{4.24}$$

With union utility given by (4.20) and the variable profits to labour costs ratio by

$$\frac{\pi_V}{\bar{w}_m L_m} = \frac{1 - \alpha\kappa}{\alpha\kappa} \tag{4.25}$$

This ratio together with the elasticity of labour demand as given by (4.19), means that equation (4.24) leads to a bargained wage of

$$w_m = \frac{\alpha\kappa + \beta(1 - \alpha\kappa)}{\alpha\kappa} \tag{4.26}$$

As the wage rate in the traditional sector has been normalised to one, equation (4.26) also characterises the wage differential between the unionised and non-unionised sector. This wage differential increases with union bargaining power β and hence reaches a maximum when unions have total bargaining power. At the other extreme, if unions have no bargaining power, the wages in the two sectors equalise irrespective of the intensity of product-market competition. Hence, even if there is imperfect competition on the product market, the reason for the wage differential between the two sectors is solely the presence of unions in the manufacturing sector. However, the conclusion that the imperfections on the goods market have no influence on the wage differential is only true for this limiting case. As can be seen from equation (4.26), since

---

[14] Unions treat the wage in the traditional sector as exogenously given. In other words, they ignore the effect wage negotiations in the manufacturing sector have on the labour supply in the traditional sector.

$\beta < 1$, the bargained wage is a decreasing function of product-market competitiveness κ. This is because more intense competition increases the elasticity of labour demand with respect to wages. Therefore, a given increase in the real wage rate will lead to a larger decrease in labour demand. In other words, the negative effect on union utility of higher wages gains in importance. For this reason, unions will lower their respective wage demands. Thus, although product-market competition is assumed to not directly influence union bargaining power, it does have an indirect effect through the elasticity of labour demand and thereby on the resulting wage rate. Therefore, not only stronger union bargaining power, but also higher product-market power of firms leads to higher relative wages.

With the wage rate in the manufacturing sector now determined, the labour market is fully specified. This makes it possible to derive the general-equilibrium results and hence to analyse the comparative statics of the model.

Only by determining the general-equilibrium values of all endogenous variables is it possible to take all direct and indirect effects of changes in exogenous variables into account. This is especially important in the model presented here, as a change in an exogenous variable which directly only affects one sector of the labour market, may well also have indirect effects on variables which influence the market outcome in the other sector as well as on the household job-taking decision. As has been shown above, due to the presence of unions in the primary sector, a positive wage differential exists between the two sectors. Since labour is treated as homogeneous in this chapter, all workers would prefer a job in the high-wage manufacturing sector. However, this wage differential leads to lower labour demand than the market-clearing level, so that not all workers can be absorbed by this sector. Therefore, workers at the beginning of their careers or those that become unemployed, must decide whether to try and obtain a high-wage job but face the risk of a period of "wait unemployment", or to enter the traditional sector where they can instantaneously find a job. In a steady-state equilibrium and with risk-neutral workers, the guaranteed income from a job in the low-wage, traditional sector, must be equal to the expected income from a high-wage job in the manufacturing sector where, however, there is also a certain probability of being unemployed and only earning unemployment benefits. Only if these two income and hence utility levels are identical, do workers not have an incentive to change their current status.

For the reasons discussed above, only workers from the unemployed pool are considered for primary jobs so that there is no direct transition between the traditional and manufacturing sector.[15] However, for the model to be economically meaningful, there must be a positive probability of finding a job in the

---

[15] As shown in van Schaik and de Groot (1998), it is easily possible to account for a direct transition from the traditional to the manufacturing sector. However, doing so only complicates the analysis below without changing the qualitative results.

unionised sector, and hence there must also be labour turnover in any period of time.[16] Therefore, it is assumed that there is an exogenous job-breakup hazard rate b similar to that first introduced in Chapter 3.3, which measures the frequency of exogenous idiosyncratic negative shocks that lead to a current job being terminated and hence to a worker becoming unemployed.[17] Further, the unemployed find a job at the rate a which is determined endogenously below. This means that all possible transitions between the states of unemployment and the primary sector are Poisson processes. As steady-state (expected) wages in both sectors are constant and with symmetrical firms whereby there is one uniform wage rate $w_m = w_M$ in the manufacturing sector, the Bellman equations for the three possible states a worker can be in, i.e. employed in the traditional sector, employed in the manufacturing sector or unemployed, are

$$\rho\tilde{V}^T = w_T \tag{4.27}$$

$$\rho\tilde{V}^M = w_M + b(\tilde{V}^U - \tilde{V}^M) \tag{4.28}$$

$$\rho\tilde{V}^U = z + a(\tilde{V}^M - \tilde{V}^U) \tag{4.29}$$

with $\tilde{V}^T, \tilde{V}^M$ and $\tilde{V}^U$ as the respective present-discounted values associated with the three states. To ensure that an individual does not have a higher lifetime utility from being unemployed, it is assumed that the imputed real return from unemployment is lower than the wage in the traditional sector.[18] Using the fact that in equilibrium the present value of becoming unemployed must be equal to that of taking up a job in the traditional sector, $\tilde{V}^U = \tilde{V}^T$, and with the manufacturing sector wage as defined by (4.26), means that equations (4.27) to (4.29) yield a job-finding rate

---

[16] Davis and Haltiwanger (1990, 1992), estimate a job-turnover rate of roughly 10% in the U.S. manufacturing sector.

[17] Recently, there has been new empirical research on unemployment scarring whereby a spell of unemployment increases the probability of future unemployment – see Gregg (2001), Arulampalam (2001), Gregory and Jukes (2001) for the UK and Flaig et al. (1993) and Muhleisen and Zimmermann (1994) for Germany. In this case, there would need to be two different job-breakup rates. However, as in the model here, all people join the primary sector from unemployment, the assumption of only one job-breakup rate is justified. Further, seeing as the empirical effects on wages of such scarring are small and only seem to be of temporary nature and the focus of this model is on the wage differential between the unionised and non-unionised sector, unemployment scarring will not be taken into account here.

[18] In the following it is assumed that this real return from unemployment is derived from income in the form of unemployment benefits. Assuming that such benefits are financed through lump-sum and hence non-distortionary taxes means that none of the steady-state results below are altered. For this reason and as is standard in a large body of the literature, we do not explicitly take taxation aspects into account.

$$a = \frac{\alpha\kappa(b+\rho)(1-z)}{\beta(1-\alpha\kappa)} \tag{4.30}$$

As can be seen from this equation, the job-finding rate increases with a higher output elasticity α. This can be explained by the fact that higher values of α which increase labour productivity and thus reduce labour costs, will lead firms to hire more labour. If, however, union bargaining power rises, then wage costs rise which forces firms to shed labour, thereby lowering the chances of finding a job in the manufacturing sector. Higher product-market competition caused by an increase in κ leads to decreasing wages and hence to lower costs so that again, firms will increase the size of their labour force. The job-finding rate also increases if the labour-turnover rate b increases, as this implies that there are more jobs that become available in any time interval. Increases in the rate of time preference mean that the future gains from finding employment in the high-wage unionised sector decrease. This will induce some workers to leave this sector and take up a job in the traditional sector. Thus, with fewer job-seekers, the job-finding rate for the remainder will rise. The opposite holds if unemployment benefits z increase. Such a rise makes becoming unemployed more attractive. At the extreme with $z = 1$, an individual has the same guaranteed income as working in the traditional sector so that all workers will become and unemployed and search for a manufacturing-sector job, meaning that the individual worker has a probability of finding such a job that tends to zero.

Equation (4.30) together with the steady-state flow condition whereby the number of workers exiting the primary sector during any period of time must equal the number of unemployed U_M finding a job in this time period, $aU_M = bL_M$, determines the number of unemployed as

$$U_M = \frac{\beta(1-\alpha\kappa)}{\alpha\kappa(b+\rho)(1-z)}bL_M \tag{4.31}$$

This equation has total employment in the manufacturing sector L_M as the only variable still to be endogenously determined.

Due to the fact that there are no institutional barriers to market entry, in a steady-state, equilibrium profits in the manufacturing sector must be equal to zero. Therefore, from the profit equation (4.17), this means that

$$p_m = \frac{w_m L_m + \tilde{F}}{m} = \frac{w_m L_m + \tilde{F}}{AL_m^\alpha} \tag{4.32}$$

must hold at all times.

Combining the optimal pricing equation (4.18) with (4.32) and inserting (4.26) for the wage rate yields

$$L_m = \frac{(\alpha\kappa)^2 \tilde{F}}{(1-\alpha\kappa)(\alpha\kappa + \beta(1-\alpha\kappa))} \tag{4.33}$$

This equation defines the per-firm employment level. As can be seen from this equation, employment is an increasing function of the fixed costs $\tilde{F}$ that a firm has. Although this result may seem counterintuitive at first, as will be shown below, higher fixed costs lead to fewer but larger firms operating in the market, i.e. output and hence labour demand per-firm rise. An increase in labour's output elasticity α also leads to a rise in per-firm labour demand, as it leads to higher output levels per employee and hence lower average unit costs. If the goods become less differentiated so that competition intensity increases (i.e. κ rises), then, as can be seen from equation (4.18), although firms will be forced to reduce their price markup, wage levels also fall and the lower prices lead to higher product demand and hence firms will hire additional workers. Higher union bargaining power, however, leads to higher wages and thus increased labour costs, so that the firm will be forced to reduce the size of its staff.

In order to determine total employment in the manufacturing sector, it is first necessary to derive the number of firms operating in this sector. Using the equilibrium household demand conditions (4.9) and (4.10) as well as the definition of the size and distribution of the population across the sectors (4.1) and (4.2) means it is possible to derive this number as[19]

$$
n = \frac{(b+\rho)(1-z)(1-\alpha\kappa)(\alpha\kappa + \beta(1-\alpha\kappa))\zeta\bar{L}}{\tilde{F}\left[(b+\rho)(1-z)[(1-\zeta)(\alpha\kappa + \beta(1-\alpha\kappa)) + (\alpha\kappa)^2\zeta] + (1-\alpha\kappa)b\beta\alpha\kappa\zeta\right]}
$$
(4.34)

Applying comparative statics shows that the number of firms is decreasing in labour's output elasticity α. As shown above, an increase in this parameter leads to firms' output levels rising. However, this higher output will lead to falling prices so that revenue falls. As can be seen from equation (4.12) and (4.13), a lower number of firms which increases the price index p_M, shifts the demand curve for the remaining firms upwards. Hence, in order to be able to cover the fixed costs, the new equilibrium is characterised by fewer but larger firms.

The effects of higher union bargaining power on the number of firms is ambiguous. If unemployment benefits are "low", then higher bargaining power leading to higher wages will lead firms to dismiss some of their workforce in order to reduce their wage bill. These lower costs imply that they earn short-run profits which will induce further firms to enter the market. If, however, unemployment benefits z are "high", then higher bargaining power leads to a reduction in the number of firms. In this case, many of the workers who are dismissed as a result of the increase in the wage level will opt to become unemployed as the income loss relative to the wage that could be earned in the traditional sector is now not so extreme. With many people searching for a job in the unionised sector and not working in the traditional sector and in

---

[19] See Appendix A.6 for a more detailed derivation

addition fewer workers now employed in the high-wage manufacturing sector, total income and hence total demand for manufacturing goods and hence the price of the these goods decreases. This means that some firms will no longer be able to cover their fixed costs and hence are forced to exit the market.

For the case that competition becomes more intense, i.e. κ rises, the effect on the number of firms will be negative. The reason for this is that as the goods become closer substitutes, firms are forced to reduce their price-markup factor. This leads to lower revenue and forces some firms out of the market with the remaining firms increasing their output in order to cover their costs. If preferences change so that ζ increases and the income share spent on manufacturing goods rises, then the rise in demand for these products and the corresponding increase in revenue means that more firms will enter the market. A rise in demand also occurs if the rate of time preference ρ increases, causing households to increase their consumption or if the size of the population $\bar{L}$ increases. In all cases, the rise in demand causes the number of firms to also increase. The number of firms decreases, however, if the job-destruction rate b or imputed income z increases as in both cases unemployment will rise leading to lower product demand.

With the equilibrium number of firms derived, it is possible to determine total labour demand and supply in the manufacturing sector, and hence the aggregate unemployment rate, i.e. the number of unemployed as a fraction of the total size of the labour force in the economy, as

$$u=\frac{b\alpha\beta\kappa\zeta(1-\alpha\kappa)}{b\alpha\beta\kappa\zeta(1-\alpha\kappa)+(b+\rho)(1-z)((1-\alpha\kappa)(\beta(1-\zeta)-\alpha\kappa\zeta)+\alpha\kappa)} \quad (4.35)$$

From this equation it can be seen, that the effect of both labour's output elasticity α as well as the intensity of competition on the product market κ on the unemployment rate is ambiguous. On the one hand, increases in either of these variables leads unions to reduce their wage demands and hence to a fall in the manufacturing-sector wage. This will induce some workers to stop searching for a high-wage job and hence reduce the number of unemployed. On the other hand, the reduction in wage costs also means that firms can reduce their price so that their product and hence labour demand increases so that the job-finding rate also rises. With the expected unemployment duration falling, workers in the traditional sector now have an incentive to join the pool of unemployed. This second effect is more likely to dominate the higher competition-intensity is, as in this case, not only do prices fall due to lower bargained wage rates, but also because firms are forced to reduce the markup over their costs, which leads to a further increase in product and labour demand. Thus, unless κ is "low", the unemployment rate falls with higher labour-output elasticity or rising product-market-competition intensity. Put differently, the more product-market power that firms possess, the higher is the unemployment rate. However, the unemployment rate is also increasing with the bargaining power that unions possess, as a higher value of β will lead to an increase in wage costs and hence force firms to dismiss some of their staff. Further, the higher wages will also entice some workers formerly in the traditional sector to now search for a job in the high-wage sector. Although this effect is slightly counteracted by the fact that the job-finding rate a decreases with higher levels of union bargaining power, the resulting effect on the unemployment rate remains unambiguously positive.

If households spend a higher income share ζ on manufacturing goods, then this will lead to more manufacturing-sector firms and hence to more demand for workers in this sector. However, with a constant job-breakup rate, the higher number of workers in this sector also means that the inflow into unemployment increases. Further, as can be seen from the job-finding rate (4.30), the rate at which workers leave unemployment remains unaltered so that the new flow equilibrium can only be achieved at a higher unemployment rate.

Increases in the rate of time preference ρ will lead to a lower unemployment rate. This can be explained by the fact that the gains in utility from searching for a high-wage job decline, as future income is valued less in present terms. Hence, some of the unemployed will now take up positions in the traditional sector. Further, a higher value of the rate of time preference also leads to households increasing their consumption level, so that the demand for goods and hence also for labour will increase which will also work towards reducing the unemployment rate.

Higher values of the job-destruction rate b as well as unemployment benefits z both lead to a rise in the unemployment rate as both increase the inflow into unemployment. In the first case, firms dismiss more workers in any given time interval and in the second case, more workers from the traditional sector will start searching for a high-wage job as higher values of this income increase the present-discounted value of becoming unemployed. This increase in utility is slightly counteracted by the now lower job-finding rate.

These and all other comparative static results are presented as an overview in Table 4.1. As can be seen from this table, not all changes in either α, β or

Table 4.1: Comparative Static Results with Union Wage Bargaining

	α	β	κ	ζ	ρ	b	z	A	$\check{F}$	$\bar{L}$
w_M	-	+	-	0	0	0	0	0	0	0
L_m	+	-	+	0	0	0	0	0	+	0
p_m	-	+	-	0	0	0	0	-	+	0
m	+	-	+	0	0	0	0	+	+	0
a	+	-	+	0	+	+	-	0	0	0
n	-	±	-	+	+	-	-	0	-	+
M	±	±	±	+	+	-	-	+	-	+
L_M	+	-	+	+	+	-	-	0	0	+
$\bar{L}_M$	±	±	±	+	-	+	+	0	0	+
p_M	±	+	±	-	-	+	+	-	+	-
u	±	+	±	+	-	+	+	0	0	0

κ lead to unambiguous effects. For example, a higher output elasticity α and higher intensity of product-market competition κ both lead to lower wages, higher employment per firm, a higher job-finding rate and to a reduction in the number of firms operating. Hence, there are market forces which will lead to an increase and others which lead to a decrease in both manufacturing output M and labour supply $\bar{L}_M$ in the high-wage sector. Similarly, the lower wage and price level associated with higher levels of α or κ will lead to a decline in the price index. However, the reduction in the number of firms leads to higher levels for this index. If union bargaining power β increases, then, as shown above, the effect on the number of firms is ambiguous. It is this ambiguity that also explains why an increase in β can lead both to a rise or fall in total manufacturing output, although under "plausible" parameter constellations, it will lead to a lower total output. Similarly, due to the ambiguous effect on the number of firms, the effects of higher bargaining power on total labour supply in the manufacturing sector can also not be precisely determined.

As was shown above, if ζ increases reflecting a change in households' preferences towards consuming more manufacturing goods or if ρ rises increasing the demand for all goods, then this will lead to a rise in total manufacturing output M and also lead to higher employment levels L_M and attract additional workers in this sector, so that $\bar{L}_M$ also rises. Further, the increase in the number of firms due to a higher value of ζ also leads to a fall in the price index as the greater variety positively enters the utility function and hence reduces the minimum costs of purchasing a unit of the composite good.

The effects of an increase in the job-breakup rate b on the job-finding rate, the number of firms and the unemployment rate have been discussed above. The higher layoff rate leads to a fall in the size of the workforce L_M and hence to a fall in manufacturing output M. With the unemployment rate rising, it also becomes less attractive to search for a job in this sector, so that some workers will now take up a job in the traditional sector so that $\bar{L}_M$ falls. Further, with the number of firms decreasing, the price index p_M will rise.

That a higher value of imputed real income z lowers the number of firms and hence leads to a rise in the price index and unemployment rate has been explained above. The reduction in the number of firms also means that labour demand L_M will fall. However, with the utility from being unemployed rising, the total number of individuals in the manufacturing sector $\bar{L}_M$ will increase.

The effects of a higher technological parameter A can all be intuitively explained. Firstly, a higher technological level will reduce firms' costs so that both the per-firm price p_m and consequently also the price index p_M will fall. Secondly, a better technology will also increase worker productivity, so that output will also rise in this sector.

As explained above, higher fixed costs $\tilde{F}$ lead to a higher levels of employment in each firm but lower the number of firms. This second effect outweighs the first, so that total output falls. Further, both the lower number of firms and the higher costs leading to a higher firm price, lead to a rise in the manufacturing price index p_M.

Finally, an increase in the size of the population $\bar{L}$ increases the demand for consumption goods. This leads to rise in the number of firms (a fall in the price index), production and employment. With more people now involved in the production process in the manufacturing sector, the (absolute) number of individuals in this sector $\bar{L}_M$ also increases.

4.2.1.2 Discussion

This section has developed a dual labour-market model which combines union wage bargaining with monopolistically competitive firms. The model predicts that increased product-market competition will lead to fewer but larger firms. It was further shown that unemployment not only positively depends on the level of unemployment benefits and on higher union bargaining power, but for very general and plausible conditions, also on the amount of market power that firms possess on the product market. There are two effects by which the level of product-market competition influences unemployment. Firstly, there is the direct effect by which more intense competition leads to lower markup prices. This price reduction is concomitant with higher product-market (and therefore labour) demand. There is, however, also an indirect effect associated with the competition intensity. More intense competition means that firms are less able to pass on increased costs to consumers. Therefore, firms must reduce their costs by dismissing a part of their workforce. Thus, higher product-market competition intensity also leads to a higher (absolute) elasticity of labour demand. However, seeing as the level of employment is important for union utility, unions will lower their wage demands correspondingly so that the unemployment rate falls. As shown in Chapter 2, this theoretical prediction is also confirmed by empirical evidence in a vast number of industries and countries. In fact, these unemployment-reducing effects of increased competition are often stated as one of the benefits of European Monetary Union. As price levels become more comparable within the EU, it does indeed seem likely that this will increase the elasticity of substitution between two variants leading not only to lower wages, but with time also to lower unemployment.

Of course, as the above model has shown, the unemployment rate can also be reduced by lowering the amount of unemployment benefits as this reduces the incentive to spend time unemployed searching for a high-wage job. However, the fact that there are social, political and normative aspects which prevent unemployment benefits from being radically reduced, emphasises that it is increasingly important that government policies designed to reduce unemployment are aimed at strengthening product-market competition and that it is recognised that unions as such only have a dwindling influence on national unemployment levels.

As explained in Chapter 3.2, firms also have an intrinsic interest in not letting wages fall to market-clearing levels and this effect alone can explain the existence of unemployment. How efficiency-wage considerations interact in a dual labour-market model with an imperfect product market is the subject of the next section.

4.2.2 Efficiency Wages

As discussed in the previous section, there is a vast body of empirical evidence which shows that there are permanent wage differences between sectors amongst observationally identical workers. Further, empirical evidence also shows that it is the existence of efficiency wages in certain sectors which may explain these differentials.[20] In order to show this more formally, the model from the previous section will be modified by assuming that wages are now not set by negotiations with unions, but instead by profit-maximising firms who are unilaterally able to set both wages and employment, but who recognise that their workforce is only prepared to provide effort if they receive an efficiency wage.[21] Thus, both the household optimisation decisions and the product-market conditions are identical to those derived in Chapter 4.1, i.e. the manufacturing sector is characterised by monopolistic competition.

The efficiency-wage considerations here are the same as in Chapter 3.2, i.e. the underlying hypothesis is that workers will only provide effort if they perceive their wage to be "fair".[22] There are several reasons why efficiency wages are assumed to only operate in the manufacturing sector. Firstly, as profits are only made in this sector, workers may find it "fair" that these profits are shared with them in the form of higher wages. Secondly, if the type of work in this sector is more complex and requires managerial skills, there may be monitoring and motivation problems which only occur in this sector.

4.2.2.1 The Model

To incorporate the fact that the amount of labour supply now needs to be measured in efficiency units, the production function as originally given by

---

[20] See, for example, Bulow and Summers (1986), Krueger and Summers (1988) and more recently Gibbons and Katz (1992) and Gera and Grenier (1994). See Albert and Meckl (2001) for an efficiency-wage model with inter-sectoral wage differentials in an open-economy.

[21] See, for example, Pichler (1993), Altenburg and Straub (1998) or Wapler (2001) for models which incorporate union wage bargaining and efficiency wages simultaneously.

[22] See Amable and Gatti (2002) for a model which combines monopolistic competition with no-shirking efficiency wages.

(4.16) needs to be modified to

$$m_i = A(\tilde{e}L_{m_i})^\alpha \tag{4.36}$$

with the effort function given by

$$\tilde{e} = -\varepsilon_1 + \varepsilon_2 \left(\frac{w_{m_i}}{w_T} \right)^{\varepsilon_3} \qquad \varepsilon_1 > \varepsilon_2(1 - \varepsilon_3),\ 0 < \varepsilon_3 < 1 \tag{4.37}$$

which is identical to the function in Chapter 3.2 with the exception that the reference income is now the wage that workers could earn in the fully competitive traditional sector.

Profits in the manufacturing sector are given as above by

$$\pi_{m_i} = p_{m_i}(m_i)m_i - w_{m_i}L_{m_i} - \tilde{F} \tag{4.38}$$

Profit-maximising firms can set both wages and employment to optimal levels. Hence, assuming symmetrical firms so that the firm index i can be dropped leads to

$$\frac{\partial \pi_m}{\partial w_m} = \frac{\partial m}{\partial \tilde{e}} \frac{\partial \tilde{e}}{\partial w_m} p_m \kappa - L_m \overset{!}{=} 0 \tag{4.39}$$

$$\frac{\partial \pi_m}{\partial L_m} = \frac{\partial m}{\partial L_m} p_m \kappa - w_m \overset{!}{=} 0 \tag{4.40}$$

It can be seen from (4.40) that, due to monopolistic competition, firms are able to charge a constant markup over their marginal costs

$$p_m = \frac{1}{\kappa} \frac{w_m}{\partial m / \partial L_m} \tag{4.41}$$

Combining equations (4.39) and (4.40) again yields the Solow condition which can be used to derive a constant optimal effort level as

$$\tilde{e}^* = \frac{\varepsilon_1 \varepsilon_3}{1 - \varepsilon_3} \tag{4.42}$$

Noting that the wage rate in the traditional sector w_T is normalised to one, this leads to an equilibrium manufacturing sector wage

$$w_m = w_M = \left(\frac{\varepsilon_1}{\varepsilon_2(1 - \varepsilon_3)} \right)^{\frac{1}{\varepsilon_3}} \tag{4.43}$$

The parameter restrictions noted in equation (4.37) ensure that the wage differential between the manufacturing and traditional sectors is always positive. As can be seen, the higher is the negative intercept of the effort function ε_1, the higher is the disutility associated with working and hence the equilibrium wage to guarantee a positive amount of effort must also be higher. Similarly, as c.p. a higher value of the parameter ε_2 leads to higher effort, an increase in this parameter means that firms can obtain the same effort level at a lower equilibrium efficiency wage. The effects of an increase in ε_3 cannot be unambiguously determined. For initially "small" values of this parameter, an increase will only lead to a small rise in effort so that the wage can be reduced slightly. However, for initially "large" values of ε_3, as can be seen from (4.37), any wage increase will lead to a relatively large increase in the amount of effort a worker provides, so that the marginal gains from increasing the wage paid to employees increases, so that in this case firms will increase the wage they pay. Finally, inspection of (4.43) shows that this wage rate is independent of the product-market power that firms possess. The reason for this result is that firms can now optimally choose the values of both the wage and employment levels and these are solely determined by the production technology and the effort supply function, both of which are independent of the intensity of product-market competition κ. However, seeing as by (4.41) the price firms charge does depend on the product-market power firms possess, this variable will still be of influence when it comes to determining the general-equilibrium values.

The employment level in each firm can again be determined by combining the optimal labour demand condition now given by (4.40) with the zero-profit condition which yields

$$L_m = \frac{\alpha \kappa \tilde{F}}{(1 - \alpha \kappa) \left(\frac{\varepsilon_1}{\varepsilon_2 (1 - \varepsilon_3)} \right)^{\frac{1}{\varepsilon_3}}} \tag{4.44}$$

As can be seen from this equation, higher values of α, κ and $\tilde{F}$ all lead c.p. to an increase in per-firm labour demand. The effects of $\varepsilon_1, \varepsilon_2$ and ε_3 on labour demand depends on how the these parameters affect the wage rate. Hence, for example, an increase in ε_1 which leads to a higher efficiency wage will induce a fall in labour demand as wage costs rise.

In order to be able to derive the unemployment rate, it is first necessary to determine the number of firms operating in the manufacturing sector and hence the job-finding rate at which the unemployed find jobs in the manufacturing sector. As the Bellman equations given by equations (4.27) to (4.29) are all independent of the type of wage setting, they all hold here as well. However, with the wage rate now defined by (4.43), the job-finding rate becomes

$$a = \frac{(b + \rho)(1 - z)}{\left(\frac{\varepsilon_1}{\varepsilon_2 (1 - \varepsilon_3)} \right)^{\frac{1}{\varepsilon_3}} - 1} \tag{4.45}$$

and hence in a steady-state flow equilibrium the number of unemployed are given by

$$U_M = \frac{\left(\frac{\varepsilon_1}{\varepsilon_2(1-\varepsilon_3)}\right)^{\frac{1}{\varepsilon_3}} - 1}{(b+\rho)(1-z)} bL_M \tag{4.46}$$

Using this result and by an analogous calculation to the one in Chapter 4.2.1.1, yields an equilibrium number of firms given by

$$n = \frac{(b+\rho)(1-z)(1-\alpha\kappa)\zeta\bar{L}}{\tilde{F}\left[(b+\rho)(1-z)(1-\zeta)+b\alpha\kappa\zeta+\alpha\kappa\zeta[(1-z)\rho-bz]\left(\frac{\varepsilon_1}{\varepsilon_2(1-\varepsilon_3)}\right)^{-\frac{1}{\varepsilon_3}}\right]} \tag{4.47}$$

Applying comparative statics to this equation yields similar results to those with union wage setting so that only new effects occur due to the effort-function parameters will be discussed here. However, changes in these variables yield ambiguous results which, as can be seen from equation (4.47) depend crucially on the value of unemployment benefits z. If these are "high" then waiting for a high-wage job is attractive and the unemployment rate is correspondingly high. In this case, a change in any of the parameters $\varepsilon_1, \varepsilon_2$ or ε_3 leading to a rise in the wage will induce even more workers to quit the traditional sector thereby leading to a further fall in aggregate demand which outweighs the rise in income for the (relatively few) workers receiving the higher wage. Hence, in this case, the number of firms will decrease. An opposite line of reasoning holds if unemployment benefits are "low".

With the number of firms determined, it is possible to derive the unemployment rate as

$$u = \frac{b\left[\left(\frac{\varepsilon_1}{\varepsilon_2(1-\varepsilon_3)}\right)^{\frac{1}{\varepsilon_3}} - 1\right]}{\left(\frac{\varepsilon_1}{\varepsilon_2(1-\varepsilon_3)}\right)^{\frac{1}{\varepsilon_3}} \frac{(b+\rho)(1-z)(1-\zeta)+b\alpha\kappa\zeta}{\alpha\kappa\zeta} - bz + \rho(1-z)} \tag{4.48}$$

It can again be seen from this equation that if unemployment benefits z are equal to unity, in other words are just as high as the wage in the traditional sector, that all workers will try and find a job in the high-wage sector and the unemployment rate reaches its maximum. Further, as the wage rate is independent of labour's output elasticity α and product-market conditions κ, changes in these variables do not influence the job-finding rate. However, increases in α which lead to lower unit costs or κ which lead to a lower markup both increase firms' demand for labour. This implies that the number of employees working in the manufacturing sector and becoming unemployed each period also increases. Hence, with the inflow into unemployment increasing but the job-finding rate remaining constant, flow equilibrium can only be reached by the unemployment rate increasing. In addition, with manufacturing-sector firms increasing their employment, the benefits of leaving the traditional sector and becoming unemployed to search for a high-wage job increase. The same line of reasoning also applies to increases in the income share ζ that households spend on manufacturing goods.

The effects of the effort-function parameters on the unemployment rate work in the same direction as their respective effects on the wage rate. The effects of the job-breakup rate b and unemployment benefits z are identical to the case with union wage setting and will therefore not be discussed here. Finally, equation (4.48) shows that the unemployment rate is an increasing function of the intensity of product-market competition κ. The reason for this is that with efficiency-wage setting product and hence labour demand increase as firms are forced to lower their prices. Although, as can be seen from equation (4.45), this does not affect the rate at which workers exit unemployment, total employment in this sector increases with more intense competition which corresponds to the empirical evidence as shown in Fig. 2.8. Therefore, in any time interval more workers will be hired so that the expected income and hence utility of choosing to become unemployed increase. Hence, the size and importance of the manufacturing sector relative to that of the traditional sector increases. Hence more workers will shift from the fully competitive sector to the sector characterised by labour-market imperfections so that the aggregate unemployment rate will rise.

All comparative static results are presented as an overview in Table 4.2. A

Table 4.2: Comparative Static Results with Efficiency-Wage Setting

	α	κ	ζ	ρ	ε_1	ε_2	ε_3	b	z	A	F	$\tilde{L}$
$\tilde{e}$	0	0	0	0	+	0	+	0	0	0	0	0
w_M	0	0	0	0	+	-	±	0	0	0	0	0
L_m	+	+	0	0	-	+	±	0	0	0	+	0
p_m	±	-	0	0	+	-	-	0	0	-	+	0
m	+	+	0	0	-	+	+	0	0	+	+	0
a	0	0	0	+	-	+	±	+	-	0	0	0
n	-	-	+	+	±	±	±	-	-	0	-	+
M	±	±	+	+	-	+	+	-	-	+	-	+
L_M	+	+	+	+	-	+	±	-	-			+
$\tilde{L}_M$	+	+	+	-	±	±	±	+	+	0	0	+
p_M	±	±	-	-	+	-	-	+	+	-	+	-
u	+	+	+	-	+	-	±	+	+	0	0	0

comparison of the results obtained by efficiency-wage setting with those by union wage setting listed in Table 4.1 shows that nearly all qualitative effects are identical. As the wage rate is now independent of labour's output elasticity α, the effects of an increase in this variable on the price a firm charges are now no longer unambiguous. On the one hand, the higher output elasticity will reduce marginal costs and hence the price level. On the other hand, the higher productivity will also lead to an increase in the number of employees per firm which increases total wage costs. If competition is very intense so that firms can only charge a small markup, then this second effect dominates and the price level needs to increase in order for firms to be able to cover their fixed costs. This ambiguity also holds for the effects of an increase on α on total manufacturing output M. Although an increase in this parameter leads to fewer but larger firms as before, for the case that the increase in the price level dominates, demand for the composite good will fall. However, in contrast to union wage setting, the total number of individuals in the manufacturing sector $\bar{L}_M$ will now unambiguously rise when α increases. The reason for this is that with union wage setting, a higher value of α led to a higher job-finding rate a which reduced the (absolute) number of unemployed and hence the number of individuals in the manufacturing sector. With efficiency-wage setting, the job-finding rate is independent of labours output elasticity and hence there is no counteracting force which can potentially lead to a lower value of $\bar{L}_M$. An analogous line of reasoning holds for the now unambiguous increase in the number of individuals in the manufacturing sector when competition intensity as measured by κ increases.

4.2.2.2 Discussion

As has been shown above, most qualitative results in a dual labour-market model with efficiency-wage setting are identical to the case when wages are set through bilateral bargaining between unions and employers. Hence, for example, the effect that more intense product-market competition leading to lower profit-margins forces some firms out of the market with the remaining firms increasing their production also holds here. However, in contrast to the model with union wage setting, the degree of competition now has no influence on the wage rate which only depends on the parameters of the effort function. This in turn means that the job-finding rate too, is independent of the intensity of product-market competition in the manufacturing sector. However, due to the negative effects of more intense product-market competition on manufacturing-sector prices and hence positive effects on product demand for these goods, there are indirect effects on the unemployment rate. The increased product demand when competition becomes more intense means that labour demand also increases and hence that the sectoral employment rate increases. Therefore, although the rate at which the individual worker finds a job out of unemployment does not change, the higher labour demand level means that more workers are able to find jobs in the manufacturing sector at the current wage. Therefore, the expected gains of becoming unemployed increase and some workers currently employed in the traditional sector will now opt for unemployment so that the aggregate unemployment rate increases as the relative importance of the imperfectly competitive sector of the labour market increases relative to that of the fully competitive traditional sector.

Further, as with union wage setting, if the government reduces the level of unemployment benefits z, then the aggregate unemployment rate decreases with workers instead taking up jobs in the low-wage traditional sector. However, here the same criticism as in Chapter 3.2.3 holds, i.e. this result rests in part on the assumption that both sides of the labour market have perfect

information and that positive insurance effects of unemployment benefits are
not taken into account.

As has been shown in this chapter so far, the degree of monopolistic competition has direct as well as indirect effects on the employment level. Further,
it also has a decisive influence on the number and size of the firms and in
the case of union wage setting, also on the wage rate that unions demand.
Irrespective of how wages are set, imperfect product-market competition en-
ables firms to charge a markup over their marginal costs. This means that
firms make super-normal unit-profits and hence that the per-worker profits
increase which in turn leads firms to change the optimal size of their workforce. How this change in profit-level influences wages and employment if the
labour market is characterised by frictions is analysed in the next section.

4.2.3 Matching Processes

The focus of this section is how imperfectly competitive product markets
interact with a primary labour market that is characterised by matching frictions. Thus, in contrast to the traditional sector containing simple, menial,
ubiquitous and universally available jobs which therefore has a fully competitive labour market, there is imperfect information about the location and
arrival time of jobs in the manufacturing sector. As above, the household and
producer decisions are identical to those throughout this chapter so that all
results from Chapter 4.1 continue to hold. Finally, as efficiency-wage considerations are not important for the analysis below, the production function in
the manufacturing sector does not need to be measured in efficiency units so
that it is again given by (4.16).

4.2.3.1 The Model

The matching frictions in the manufacturing sector are similar to those discussed in Chapter 3.3.[23] Hence, the matching function is again of the functional form

$$\tilde{m} = \tilde{m}(u, v) \tag{4.49}$$

---

[23] See Davidson et al. (1988) for a dual labour-market model with matching frictions in one sector of the economy and heterogeneous labour. However, in their
equivalent to the manufacturing sector here, a Leontief production technology is
assumed. A different setup is chosen by Acemoglu (2001b). In his model, one final
good is produced using two immediate goods as inputs where each input is produced in a separate sector. Matching frictions are present in both these sectors,
but due to different production technologies, there is a positive wage differential
between the two sectors. In contrast to here, both these models assume perfect
competition in the product markets.

which means that the rate at which firms fill vacancies an workers find jobs are also identical to those derived in Chapter 3.3 and are given by $f(\theta)$ and $f(\theta)\theta$ respectively, where θ is again an indicator of labour-market tightness from the employer perspective.

In the matching model of the previous chapter, labour productivity was exogenously fixed. With the production technology as given by equation (4.16), this is no longer the case here. Instead, firms face a dynamic optimisation problem in optimally choosing their number of vacancies subject to employment fluctuations within the firm. This means that the relevant Hamiltonian function can be expressed as

$$\mathcal{H} = e^{-rt}\left[p_{m_i}m_i - w_{m_i}L_{m_i} - cA\Upsilon_i - \tilde{F}\right] + \Lambda[f(\theta_i)\Upsilon_i - bL_i] \qquad (4.50)$$

where Υ_i denote the number of vacancies of firm i. Dynamic optimisation leads to

$$\mathcal{H}_{\Upsilon_i} = -e^{-rt}cA + \Lambda f(\theta_i) \overset{!}{=} 0$$

$$\mathcal{H}_{L_i} = e^{-rt}\left[\frac{\partial p_{m_i}}{\partial m_i}\frac{\partial m_i}{\partial L_i}m_i + p_{m_i}\frac{\partial m_i}{\partial L_{m_i}} - w_{m_i}\right] - b\Lambda = -\dot{\Lambda}$$

from which the firm job-creation condition is derived as

$$\kappa\tilde{y}_i - w_{m_i} = \frac{r+b}{f(\theta)}cA \qquad (4.51)$$

where $\tilde{y}_i \equiv p_i\frac{\partial m_i}{\partial L_{m_i}}$ is the marginal value of labour output. This condition is identical to the job-creation condition (3.33) with the exception that productivity is now no longer constant and that the degree of product-market competition also influences the revenues from a position.

In contrast to the previous models in this chapter where workers and firms could instantaneously join and part so that only per-period profits are of importance, with matching frictions, search and hiring costs are incurred up front so that the duration of a match is decisive, i.e. the expected gains in future periods also need to be considered as they need to be high enough to cover the initial costs. Thus, the Bellman equations (3.30), (3.32), (3.34) and (3.35) giving the present-discounted values for firms and workers derived in Chapter 3.3 all continue to hold – with the exception that the productivity of a position needs to be endogenously determined – and are simply repeated here for convenience

$$rW^V = f(\theta)(W^F - W^V) - c\tilde{y} \qquad (4.52)$$
$$rW^F = \tilde{y} - w_m - bW^F \qquad (4.53)$$
$$r\tilde{V}^E = w_m - b(\tilde{V}^E - \tilde{V}^U) \qquad (4.54)$$

$$r\tilde{V}^U = z + f(\theta)\theta(\tilde{V}^E - \tilde{V}^U) \tag{4.55}$$

As these equations are the same as in the previous chapter, and still assuming that the wage rate is derived from a Nash bargaining solution, means that in analogy to equation (3.38), the wage rate is given by

$$w_{m_i} = \tilde{\beta}\tilde{y}_i + (1 - \tilde{\beta})r\tilde{V}^U \tag{4.56}$$

However, in contrast to the model from the previous sector which only analysed one sector of the economy, here, with risk-neutral workers seeking to equalise expected utility from either joining the traditional or manufacturing sector, the returns to choosing unemployment must equal those from taking up a job in the traditional sector, i.e. $\tilde{V}^U = \tilde{V}^T$. Noting from the Keynes–Ramsey rule (4.8) that $r = \rho$ holds in equilibrium, means from the Bellman equation for a worker in the traditional sector (4.27) that $r\tilde{V}^U = w_T = 1$ so that the above wage equation simplifies to

$$w_m = w_M = \tilde{\beta}\tilde{y} + (1 - \tilde{\beta}) \tag{4.57}$$

where the firm index has been omitted due to the assumption of symmetrical firms. In order to be able to explicitly determine the wage, it is first necessary to derive the value of marginal output $\tilde{y}$.

With monopolistic competition, the pricing markup as given by (4.18) continues to hold. As hiring costs are incurred before production starts (and need to be covered by the expected share of the surplus accruing to the firm), the zero-profit condition (4.32) still holds. Equating these two equations leads to a firm labour demand given by

$$L_m = \frac{\alpha\kappa\tilde{F}}{w_m(1 - \alpha\kappa)} \tag{4.58}$$

Using this result to derive an expression for $\tilde{y}$ results in

$$\tilde{y} = p_m \frac{\partial m}{\partial L_m} = \frac{w_m L_m + \tilde{F}}{A L_m^\alpha} A\alpha L_m^{\alpha-1}$$
$$= \frac{\alpha(w_m L_m + \tilde{F})}{L_m}$$

which by inserting (4.58) becomes

$$\tilde{y} = \alpha w_m + \frac{w_m(1 - \alpha\kappa)}{\alpha\kappa}$$
$$= w_m \left(\alpha + \frac{1 - \alpha\kappa}{\alpha\kappa} \right) \tag{4.59}$$

Inserting this result into the wage equation (4.57) yields a wage rate of

$$w_M = \frac{\alpha\kappa(1 - \tilde{\beta})}{\alpha\kappa(1 + \tilde{\beta}(1 - \alpha)) - \tilde{\beta}} \tag{4.60}$$

For this wage to be larger than or equal to 1 it must hold that

$$\alpha\kappa(1 - \tilde{\beta}) \geq \alpha\kappa(1 + \tilde{\beta}(1 - \alpha)) - \tilde{\beta}$$

which simplifies to

$$\tilde{\beta}(1 - \alpha\kappa(2 - \alpha)) \geq 0$$

so that both $\tilde{\beta} \geq 0$ and

$$\kappa \leq \frac{1}{\alpha(2 - \alpha)}$$

must hold, which are both always fulfilled. However, to guarantee that a positive wage results, the numerator must be positive which means that

$$\tilde{\beta} < \frac{\alpha\kappa}{1 - \alpha\kappa(1 - \alpha)} \tag{4.61}$$

i.e. if the bargaining power is too high, the wage would be higher than the value of labour's marginal productivity so that no jobs in this sector would be created. Note further that if workers have no bargaining power, the wage rate in this sector exactly equals the wage in the traditional sector.

If the condition for $\tilde{\beta}$ as given by (4.61) is fulfilled, then comparative statics yield

$$\frac{\partial w_m}{\partial \alpha} < 0, \frac{\partial w_m}{\partial \tilde{\beta}} > 0, \frac{\partial w_m}{\partial \kappa} < 0$$

Hence, it can be seen that an increase in labour's output elasticity will induce firms to hire additional workers so that the marginal output of the individual worker falls. That the wage increases with higher worker bargaining power is to be intuitively expected. Further, as a higher bargaining power leads firms to reduce the size of their staff, the marginal revenue product increases. Finally, as a higher degree of product-market competition reduces the price markup, it also reduces the value of the surplus created by a filled position so that the nominal and real wage must fall as a result.

Inserting the wage rate into the value of labour's marginal productivity as given by (4.59) yields

$$\tilde{y} = \frac{(1 - \alpha\kappa(1 - \alpha))(1 - \tilde{\beta})}{\alpha\kappa(1 + \tilde{\beta}(1 - \alpha)) - \tilde{\beta}} \tag{4.62}$$

from which follows

$$\frac{\partial \tilde{y}}{\partial \alpha} < 0, \frac{\partial \tilde{y}}{\partial \tilde{\beta}} > 0, \frac{\partial \tilde{y}}{\partial \kappa} < 0 \qquad (4.63)$$

i.e. the same comparative statics as for the wage rate hold.

Equilibrium labour-market tightness is derived by inserting the wage rate as given by (4.60) and the value for labour's marginal value of output (4.62) into the equilibrium job-creation condition (4.51) to yield

$$\frac{(1-\alpha)(1-\alpha\kappa)(1-\tilde{\beta})\kappa}{\alpha\kappa(1+\tilde{\beta}(1-\alpha)) - \tilde{\beta}} - \frac{b+r}{f(\theta)}cA = 0$$

from which $f(\theta)$ is derived as

$$f(\theta) = \frac{(b+\rho)(\alpha\kappa(1+\tilde{\beta}(1-\alpha)) - \tilde{\beta})}{(1-\alpha)(1-\alpha\kappa)(1-\tilde{\beta})\kappa}cA \qquad (4.64)$$

Further, as the returns from taking up a traditional job or becoming unemployed must be equal, the equilibrium job-finding rate a is given from the Bellman equations by

$$a = \frac{(b+\rho)(1-z)}{w_m - 1} = f(0)0 \qquad (4.65)$$

which can be solved for $f(\theta)$ as

$$f(\theta) = \frac{(b+\rho)(1-z)(\alpha\kappa(1+\tilde{\beta}(1-\alpha)) - \tilde{\beta})}{\theta\tilde{\beta}(1-\alpha\kappa(2-\alpha))} \qquad (4.66)$$

Solving (4.64) and (4.66) for θ yields

$$\theta = \frac{(1-z)(1-\alpha)(1-\alpha\kappa)(1-\tilde{\beta})\kappa}{\tilde{\beta}(1-\alpha\kappa(2-\alpha))cA} \qquad (4.67)$$

which leads to comparative statics given by

$$\frac{\partial \theta}{\partial \alpha} < 0, \frac{\partial \theta}{\partial \tilde{\beta}} < 0, \frac{\partial \theta}{\partial \kappa} > 0, \frac{\partial \theta}{\partial c} < 0, \frac{\partial \theta}{\partial z} < 0, \frac{\partial \theta}{\partial A} < 0$$

In the case of $\tilde{\beta}, c$ and z, these results are identical to those in the previous chapter. Although a higher value of labour's marginal output elasticity α induces firms to hire more labour and hence create more vacancies, this expansion in labour demand and output leads to a rise in total wage costs and a fall in the price level so that some firms are forced to exit the market. Some of these dismissed workers will then opt to become unemployed in the hope of finding a new job in a manufacturing-sector firm. This too, has the effect of reducing labour-market tightness. If the degree of competition in the product market becomes more intense, then firms will need to reduce their prices and hence experience a fall in the surplus that accrues from a filled position. This causes some firms to exit the market. However, this effect is outweighed by the additional demand for (the now cheaper) manufacturing goods which the remaining firms now face. These need to expand their production in order to be still able to cover their fixed costs and hence they increase the number of vacancies. Further, with the wage rate falling, some of the unemployed will now take up jobs in the traditional sector which further works to increase labour-market tightness. Finally, although a higher technological level as denoted by A increases the value of marginal output, this higher productivity is transmitted into lower prices and hence does not directly change the amount of job creation. However, as hiring costs which need to be borne upfront are assumed to be proportional to this technological level, higher technology increases hiring costs and therefore indirectly reduces job creation and hence labour-market tightness.

It also needs to be noted that labour-market tightness is independent of both the job-destruction rate b and the rate of time preference ρ. The reason for this is that although both these variables influence the degree of job-creation as can be seen from equation (4.51), they also enter the equilibrium job-finding rate (4.65), i.e. job-seekers immediately adapt to this change in job-creation and leave the pool of unemployed to take up traditional-sector jobs instead.

The model is now fully specified, making it possible to again analyse the comparative static effects in the resulting general-equilibrium. Using the same derivation techniques as with union- or efficiency-wage setting, the number of firms is as

$$n = \frac{(1 - \alpha\kappa)(1 - \tilde{\beta})\zeta\theta f(\theta)\bar{L}}{\tilde{F}\left[\alpha\kappa\zeta(1 + \tilde{\beta}(1 - \alpha))(b + \theta f(\theta)) + \theta f(\theta)(1 - \tilde{\beta} - \zeta) - b\tilde{\beta}\zeta\right]} \quad (4.68)$$

where θ and $f(\theta)$ as defined above have been left in the expression for notational simplicity. The comparative static results are almost identical to the case with union wage bargaining with the exception that higher worker bargaining power now always leads to an increase in the number of firms. In the resulting equilibrium, these will be operating at smaller output levels, but due to the increase in the price of the manufacturing good associated with the increase in $\tilde{\beta}$, the revenue from the lower output is sufficient to cover the firm's fixed costs.

With the number of firms as above, the unemployment rate is derived as

$$u = \frac{b(\alpha\kappa\zeta(1 + \tilde{\beta}(1 - \alpha)) - \tilde{\beta}\zeta)}{\alpha\kappa\zeta(1 + \tilde{\beta}(1 - \alpha))(b + \theta f(\theta)) + \theta f(\theta)(1 - \tilde{\beta} - \zeta) - b\tilde{\beta}\zeta} \quad (4.69)$$

Noting from (4.67) that $\theta = 0$ for $z = 1$, it can again be seen that the unemployment rate becomes one when $z = 1$, i.e. when unemployment benefits reach the same level as taking up a job in the traditional sector. The other comparative static results are also similar to those when wages are set through union wage bargaining with the exception that higher values of labour's output elasticity α or increases in the intensity of product-market competition now both unambiguously raise the unemployment rate. Although a higher value of α lowers the wage rate, total wage costs increase as firms expand output. In the case of a higher value of κ, the resulting higher product demand causes employment in this sector to increase. This increases the present-discounted utility from searching for a job in the manufacturing sector so that workers in the traditional sector will leave their jobs and join the pool of unemployed. Thus, as with efficiency wages, the employment rate in the manufacturing sector increases with more intense competition so that there are now more workers in the imperfectly competitive sector causing the macroeconomic unemployment rate to increase. Finally, hiring costs which were not present in the union-wage setup but which lead to higher costs, also have a negative effect on the number of firms and hence a positive effect on the unemployment rate.

Table 4.3 summarises the remaining comparative static results. As can be seen,

Table 4.3: Comparative Static Results with Matching Frictions

	α	β	κ	ζ	ρ	b	c	z	A	$\tilde{F}$	$\bar{L}$
w_M	−	+	−	0	0	0	0	0	0	0	0
$\tilde{y}$	−	+	−	0	0	0	0	0	0	0	0
θ	−	−	+	0	0	0	−	−	−	0	0
L_m	+	−	+	0	0	0	0	0	0	+	0
p_m	−	+	−	0	0	0	0	0	−	+	0
m	+	−	+	0	0	0	0	0	+	+	0
a	−	−	+	0	0	0	−	−	−	0	0
n	−	+	−	+	0	−	−	−	−	−	+
M	±	−	±	+	0	−	−	−	+	−	+
L_M	+	−	+	+	0	−	−	−	−	0	+
$\bar{L}_M$	+	−	+	+	0	+	+	+	+	0	+
p_M	−	+	−	−	0	+	+	+	−	+	−
u	+	+	+	+	0	+	+	+	0	0	0

these results are very similar to those obtained when wages are set through bilateral bargaining between unions and employers. The only difference is that now, as explained above, the rate of time preference ρ does not affect θ and therefore does not influence the number of firms, manufacturing sector output and hence also leaves the number of workers in this sector unaltered. New is also the effect of hiring costs: as these increase firms costs they lead to a reduction in the number of manufacturing-sector firms and hence to a decrease in employment.

4.2.3.2 Discussion

The above has integrated a dual labour-market model with monopolistic competition on the product market and matching frictions in the high-wage sector. As the degree of competition intensity determines the price markup, it indirectly influences the surplus that accrues when a suitable job-worker pair is matched. Higher competition intensity not only reduces the wage rate because the surplus decreases, but also the markup factor so that the price for each variant further decreases. The resulting higher demand for these products means that firms' demand for labour increases. Thus, within the manufacturing sector, higher competition intensity increases the employment rate. However, within the dual labour-market setup presented here, the higher job-finding rate which results from increased product-market competition means that the expected returns to searching for a high-wage job increase. Although this effect is slightly counteracted by the fall in the real wage, some workers currently employed in the traditional sector will quit their jobs there and join the pool of unemployed. This means that the traditional sector becomes smaller relative to the manufacturing sector and hence that there is a shift away from the perfectly towards the imperfectly competitive sector so that at an aggregate level, the unemployment level rises.

4.3 Summary and Conclusions

In this chapter different types of labour-market imperfections have been separately analysed. In reality, all three kinds will of course be simultaneously present. This has important implications for potential government policies aimed at reducing unemployment as the only common policy element that all the theoretical models have in addition to policies which influence the intensity of product-market competition is the role that unemployment benefits play. Even if unemployment benefits do not directly influence the wage rate, they make the option of unemployment and searching for a high-wage job more attractive. However, as stated in the previous chapter, this unambiguously positive relationship between unemployment benefits and the unemployment rate is no longer guaranteed if further market imperfections such as insurance aspects are integrated. Therefore, irrespective of the type of wage formation, it is the imperfections on the goods market which not only have a strong influence on the performance of the labour market, but more importantly, especially for the cases when wages are either set through bilateral bargaining between autonomous unions and employers or by profit-maximising efficiency-wage considerations, government policies cannot influence the wage outcome and must therefore focus on making the product markets more transparent. Even in the case when matching frictions are the cause of labour-market imperfections, the scope for government policies on the labour market is limited to the extent that actions must be undertaken to improve the speed with which jobs and workers find each other. This can be done by improving the matching "technology", e.g. by improving the quality of job centres as is currently being discussed in Germany.

The result that the employment rate increases with product-market power holds irrespective of the wage-formation process. Further, in the case of union wage bargaining and matching frictions the wage premia declines with the intensity of competition. Both these facts correspond to the empirical evidence presented in Chapter 2. In the case of efficiency wage setting, wages are set solely be effort-supply considerations which are independent of the conditions on the product market. In the case of union wage setting, the decline in the wage rate associated with more intense competition can be strong enough to reduce the expected income associated with waiting for a job in the manufacturing sector so that in this case, the macroeconomic unemployment rate decreases. This highlights the fact that the high long-term unemployment witnessed in continental Europe cannot be solved by solely focusing on the labour market, but that imperfections in the product market have important implications for the labour-market outcome.

The models in this chapter have all integrated basic dynamic elements. Thus, households maximise their intertemporal utility and the decision whether to spend time unemployed and search for a high-wage job or not is dependent upon the rate at which future income is discounted. Further, the rate of time preference directly influences the interest rate and hence in the matching models, determine the opportunity costs of opening up a new position and hence on the job-creation decision. However, one important dynamic aspect which has been neglected so far is the fact that the production technology is not static. Therefore, the next chapter focuses on how labour-market imperfections influence the growth rate of the economy.

5

Exogenous Growth and Unemployment

Is growth "good" for employment, i.e. do higher growth levels lead to lower unemployment levels as often stated in public? Intuitively, it is not clear whether growth and unemployment are positively or negatively related as there are two main effects that need to be considered. On the one hand, one of the main channels for positive growth rates is an increase in productivity. However, if this technological progress is labour-saving, i.e less labour is required to reach the same output levels as before, then, unless this increased output is absorbed by a corresponding rise in demand, unemployment and growth will be positively linked. This also makes clear that technological progress can not only affect the aggregate (un)employment level, but for the case that the optimal factor intensities change as a result, also the functional distribution between, for example, labour and capital income.

On the other hand, if the additional income that is generated from the higher productivity levels or if new technologies that are implemented lead to reduced costs and hence lower prices, then both these factors lead to additional product demand and hence an induced increase in labour demand. If this case, there is a negative relationship between growth and unemployment. For example, Flaig and Rottmann (2001) find that for the German manufacturing sector, the critical "employment threshold", i.e. the output growth rate that is required to keep employment stable, is 0.17% in the short and 0.60% in the long run.[1]

---

[1] See, however, Buscher et al. (2000) who differentiate between an employment threshold which keeps the total labour volume constant and the threshold which keeps the unemployment rate constant. This is particularly important for Germany where labour supply is very elastic and further, increases in demand often lead to increased overtime instead of hiring new workers. Therefore, they estimate much larger values of the employment threshold of 1.85% for Germany. Further, although they find a similar value of 1.68% for Great Britain, the value for the United States is much lower and estimated as 0.50%.

How growth and unemployment are related has been the subject of intense research recently and will be the focus of both this and the next chapter. Of course, there are many mechanisms by which economic growth can occur, for example through the accumulation of human capital, investments in R&D etc. In this chapter, the source of growth is technological progress. In order to be able to solely focus on how different wage-setting mechanisms influence the relationship between the growth and unemployment rates, in this chapter the different sectors of the economy are aggregated to one all-encompassing sector and technological progress is assumed to be exogenous. Hence, the basic setup is identical to the original neoclassical model developed by Solow (1956) and Swan (1956) and extended by Cass (1965) and Koopmans (1965) to be able to endogenously determine the household consumption decision. However, in contrast to these models where the labour market is always assumed to clear, here the effects of unions, efficiency wages and matching frictions are integrated.

5.1 The Basic Model

As in Cass (1965) and Koopmans (1965), the production technology is assumed to have labour and capital as inputs, i.e.

$$Y = F(K, A_t L) \tag{5.1}$$

where Y is total output which can be either consumed or invested and K the capital stock. This function is assumed to exhibits constant returns to scale and have positive but diminishing returns with respect to each input and fulfills the Inada conditions, following Inada (1963) whereby

$$\lim_{K \to 0} \frac{\partial Y}{\partial K} = \lim_{L \to 0} \frac{\partial Y}{\partial L} = \infty$$
$$\lim_{K \to \infty} \frac{\partial Y}{\partial K} = \lim_{L \to \infty} \frac{\partial Y}{\partial L} = 0$$

As can be seen from the production function (5.1), the technology parameter A is now a function of time. Due to exogenous technological progress, this parameter is assumed to increase at the rate g_A, i.e.

$$A(t) = e^{g_A t} A_0 \tag{5.2}$$

where A_0 is an exogenously given initial productivity level. This technological progress is "disembodied" in the sense that all jobs benefit from the higher productivity without needing to replace the capital stock. Omitting the time index t for notational convenience and using the fact that the above production function is assumed to exhibit constant returns to scale, means that it is possible to rewrite equation (5.1) in intensive form as

$$\hat{y} = \hat{F}(\hat{k}) \tag{5.3}$$

with $\hat{y} \equiv Y/AL$ as per-worker output, $\hat{k} \equiv K/AL$ as the capital intensity both measured in efficiency units, and $\hat{F}(\hat{k}) \equiv F(K,1)$. From this, the marginal productivity of capital can be derived as

$$\frac{\partial Y}{\partial K} = \hat{F}'(\hat{k}) \tag{5.4}$$

and that of labour as

$$\frac{\partial Y}{\partial L} = A\big(\hat{F}(\hat{k}) - \hat{k}\hat{F}'(\hat{k})\big) \tag{5.5}$$

from which it can be seen that the marginal productivity of labour increases at the rate of technological progress.

Normalising the price of the final good to unity means that the solution to the dynamic household optimisation problem again yields the Keynes–Ramsey rule given by

$$\frac{\dot{C}}{C} = \frac{1}{\gamma}(r - \rho) \tag{5.6}$$

Using analogous methods, the intertemporal optimisation problem for a representative firm can be formulated as

$$\mathscr{H} = e^{-rt}\big(F(K, AL) - wL - I\big) + \Lambda(I - \delta K) \tag{5.7}$$

where I denotes a firm's investment, δ is the rate of capital depreciation and L and I are the control variables and K the state variable. Solving this optimisation problem yields

$$\hat{F}'(\hat{k}) - \delta = r \tag{5.8}$$

Defining $\hat{c}$ as household consumption measured in efficiency units and hence noting that

$$\frac{\dot{\hat{c}}}{\hat{c}} = \frac{\dot{C}}{C} - \frac{\dot{A}}{A} = \frac{\dot{C}}{C} - g_A$$

As the only source of growth in this economy is exogenous technological progress, the growth rate of final output equals the growth rate of technological progress, i.e. $g_A = g$. Combining this result with the Keynes–Ramsey rule and equation (5.8) gives

$$\frac{\dot{\hat{c}}}{\hat{c}} = \frac{1}{\gamma}\left(\hat{F}'(\hat{k}) - \delta - \rho - \gamma g\right) \tag{5.9}$$

as the steady-state rate of consumption growth in efficiency units. Further, as in a closed economy the representative household must end up with zero debt, the average assets per household must equal the average capital stock per household, i.e. $G = A\hat{k}$. From this it follows that

$$\dot{G} = \dot{A}\hat{k} + A\dot{\hat{k}}$$

$$\frac{\dot{G}}{A} = g\hat{k} + \dot{\hat{k}} \tag{5.10}$$

Using this, the household budget constraint (4.6) as well as equations (5.5) and (5.8) from above, and normalising the macroeconomic price index to unity yields

$$\dot{G} = rG + I_w - C$$
$$= \left(\hat{F}'(\hat{k}) - \delta\right)A\hat{k} + A\left[\hat{F}(\hat{k}) - \hat{k}\hat{F}'(\hat{k})\right] - C$$
$$= A\left(\hat{F}(\hat{k}) - \delta\hat{k}\right) - C$$

Combining this with equation (5.10) yields

$$\dot{\hat{k}} = \hat{F}(\hat{k}) - (\delta + g)\hat{k} - \hat{c} \tag{5.11}$$

In the steady-state equilibrium, household consumption and capital measured in efficiency units must be constant, i.e. $\dot{\hat{c}} = \dot{\hat{k}} = 0$. Therefore, equation (5.9) becomes

$$\hat{F}'(\hat{k}) = \delta + \rho + \gamma g \tag{5.12}$$

which uniquely determines the steady-state capital intensity.

This completes the description of the general model within a fully competitive setting. In the following, extending Ramser (1997) who analyses the effects of unions, efficiency wages and matching frictions in the context of the Solow (1956) growth model, it will be analysed how the growth and unemployment rates are related depending on the type of labour-market imperfection when consumption and savings are endogenously determined. Hence, the different forms of wage setting first introduced in Chapter 3 are integrated into the above model.

5.2 Wage Setting and General Equilibrium

5.2.1 Union Wage Bargaining

As shown above, if technological change is labour-augmenting, then in the steady-state equilibrium, the real wage increases at the rate of this technological progress. Therefore, as long as the nominal wage rate grows at the same rate as labour productivity, no employment effects will occur. However, since unions derive utility from the wage level and the size of the employed force, there are no mechanisms which guarantee that unions will find it optimal (i.e. utility maximising) to bargain for wage increases which exactly correspond to the rate of technological progress. Further, since as shown above, the level of technological progress influences the optimal capital intensity, it also has indirect effects on labour productivity and hence the wage level. Therefore, a change in the growth rate means that unions need to adapt their wage demands. These arguments will be analysed more formally below.

5.2.1.1 The Model

In the sectoral union wage-setting model in Chapter 3.1, it was assumed that there was another sector in the economy in which any worker could earn the (market-clearing) alternative wage $\bar{w}$. In Chapter 4.2.1, this alternative wage was the wage w_T paid in the traditional sector. However, in order to be able to focus on the effects of unions on growth and unemployment, the above growth model is specified with an aggregate production function. Hence, with only one sector in the economy, the union fallback position needs to be modified as now the unemployed do not have the option of taking up a job in a different sector of the economy. Instead, a dismissed worker faces a certain probability of finding a job with a different firm, or, if he is unsuccessful in this, of receiving unemployment benefits. With homogeneous workers, the probability of finding an alternative job is exactly equal to the fraction of the workforce that is employed. Therefore, the alternative income measured in efficiency units $\hat{w}_A$, is given by

$$\hat{w}_A = (1 - u)\hat{w} + u\hat{z} \tag{5.13}$$

with $\hat{w}$ denoting the wage rate and $\hat{z}$ the level of unemployment benefits measured in efficiency units respectively. Thus, the union objective function is now given by

$$V = L(\hat{w})[\nu(\hat{w}) - \nu(\hat{w}_A)] \tag{5.14}$$

With the production technology containing capital as an input factor, firms still incur capital costs if no agreement is reached with unions and hence have a negative fallback position. Therefore, with union utility as given by (5.14), the Nash bargaining problem is given by

$$\max_{\hat{w}} \Omega = \left[L[\nu(\hat{w}) - \nu(\hat{w}_A)]\right]^{\beta} \left[F(K, AL) - wL\right]^{1-\beta} \tag{5.15}$$

which again leads to an equilibrium wage $\hat{w}^*$ given by

$$\epsilon_{\nu,\hat{w}^*} = \frac{1-\beta}{\beta} \frac{\epsilon_{F,L}}{1 - \epsilon_{F,L}} - \epsilon_{L,w} \tag{5.16}$$

However, in contrast to the union models in the previous chapters, if workers and hence unions are assumed to have a certain degree of risk-aversion, i.e. $\nu''(\hat{w}) < 0$, then the elasticity of utility with respect to wages is now an increasing function of the unemployment rate. Therefore, with a concave utility function, the above equation defines a negative relationship between the wage level and unemployment rate as shown in Fig. 5.1.[2] Such a relationship is

Fig. 5.1: Wage Curve

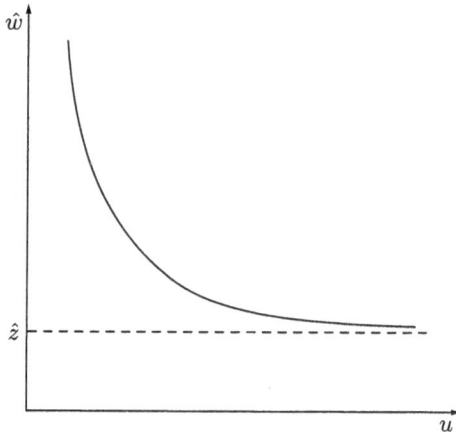

labelled the "wage curve" in the literature and has recently again become the focus of intense research.[3] Thus as can be seen from the figure, as soon as unions have a minimum degree of bargaining power and are able to bargain a wage that is higher than the level of unemployment benefits $\hat{z}$, that unemployment will result. The higher the bargaining power that unions possess, the higher will be the level of unemployment that they are willing to accept in return for a higher wage level. Further, seeing as unions and employers bargain over wages measured in efficiency units, the steady state is characterised by

---

[2] If workers are risk-neutral, then they are interested in wage-bill maximisation so that in this case, the wage curve in Fig. 5.1 becomes horizontal.

[3] See Blanchflower and Oswald (1994b) and the review by Card (1995).

a constant wage and hence unemployment level with the nominal (and real) wage increasing at the rate of technological progress.

How an increase in the rate of technological progress and hence in the macro-economic growth rate g affects the unemployment rate, can be seen from the above wage equation (5.16) whereby

$$\hat{w}^* = \hat{w}^*\left(\hat{k}^*, \beta, u, \hat{z}\right) \tag{5.17}$$

with $\partial \hat{w}^*/\partial \hat{k}^* > 0, \partial \hat{w}^*/\partial \beta > 0, \partial \hat{w}^*/\partial u < 0$ and $\partial \hat{w}^*/\partial \hat{z} > 0$. Further, as the capital market is still assumed to be perfectly competitive, the marginal productivity of capital condition as given by (5.12) continues to hold so that even in the presence of unions, the capital intensity is identical to the fully competitive labour-market case.[4] Therefore, with the labour market now not clearing, the absolute level of capital must also be lower.

With the right-to-manage approach, firms will unilaterally set the employment level L with labour demand determined by the marginal-productivity condition (5.5). Therefore, with labour demand a negative function of the wage rate, the unemployment rate is implicitly defined by

$$u = 1 - \frac{L\left(\hat{w}^*(\hat{k}^*, \beta, u, \hat{z})\right)}{\bar{L}} \tag{5.18}$$

Noting from (5.12) that the steady-state capital intensity is a (negative) function of the growth rate of technological progress, means that from the implicit-function theorem, it is possible to derive the relationship between the unemployment and growth rate as

$$\frac{\mathrm{d}u}{\mathrm{d}g} = -\frac{\frac{\partial L}{\partial \hat{k}^*}\frac{\partial \hat{k}^*}{\partial g} + \frac{\partial L}{\partial \hat{w}^*}\frac{\partial \hat{w}^*}{\partial \hat{k}^*}\frac{\partial \hat{k}^*}{\partial g}}{1 + \frac{\partial L}{\partial \hat{w}^*}\frac{\partial \hat{w}^*}{\partial u}} \tag{5.19}$$

As can be seen from equation (5.12), the steady-state capital intensity decreases with higher growth rates. Further, the wage rate is a positive function of the capital intensity because it increases labour's productivity. In addition, as from the production technology labour demand negatively depends on the wage rate and a higher unemployment rate exerts causes unions to reduce their

---

[4] See Ramser (1997) who integrates union wage bargaining into a Solow growth model. In this model without consumer-utility maximisation, the optimal capital intensity is given by the condition that savings are equal to the effective depreciation rate of the capital-labour ratio, i.e.

$$(1 - \hat{c})\hat{F}(\hat{k}) = (\delta + g)\hat{k}$$

In this case, the higher wage level associated with unions leads to capital being substituted for labour which means that the steady-state capital intensity increases with the presence of unions.

wage demands, both the numerator and denominator of the above equation are positive so that there is an unambiguously negative relationship between the unemployment and growth rates. How strong these employment growth effects are depends on the elasticities of labour demand with respect to wages, the elasticity of substitution between labour and capital and on the elasticity of wages with respect to the unemployment rate. Empirical estimates for the elasticity of labour demand with respect to wages differ greatly depending on, for example, the industry studied – see Hamermesh (1986) for a general overview and Fitzenberger (1999) for Germany. Most estimates for the elasticity of substitution between labour and capital find a value slightly below one – see Hamermesh (1993) for an overview of studies for the US and Deutsche Bundesbank (1995) who estimate a substitution elasticity of 0.8 for Germany. The values for the wage-unemployment elasticities are also low: Blanchflower and Oswald (1994a,b) estimate values of around -0.08 to -0.11 for the US, Canada and Britain and Baltagi and Blien (1998) estimate a value of -0.06 for Germany. Hence, although the relationship between and unemployment and growth is negative, its expected magnitude is fairly small.

5.2.1.2 Discussion

The above model has integrated union wage bargaining into a neoclassical growth model with intertemporal household-utility maximisation and hence an endogenous saving rate. As is to be intuitively expected, the presence of unions means that real wages do not fall to market-clearing levels so that unemployment will occur in the steady-state equilibrium. How high this unemployment rate is depends on the level of unemployment benefits and the degree of bargaining power that unions possess as these directly influence the wage rate. The rate of technological progress has an indirect effect on the wage rate as it increases the effective depreciation rate of capital and hence leads to a lower capital intensity and thereby to a lower marginal productivity of labour. However, the higher rate of (labour-augmenting) technological progress directly raises the marginal productivity of labour increases at a faster rate so that firms will want to substitute capital with labour. As this latter effect dominates, a higher steady-state growth rate of technological progress also leads to a lower equilibrium unemployment rate.

As can be seen from the union model, it is the effect of technological progress on the wage level which determines how unemployment and growth are linked. Therefore, whether this relationship also holds when wages are determined by efficiency-wage considerations, depends on how wages react to different levels of technological change and is the subject of the next section.

5.2.2 Efficiency Wages

In contrast to unions which must take the labour-demand function into account when setting their wage demands, with efficiency wages, firms are able to unilaterally set both wages and employment but must give their workforce sufficient incentives to provide effort. Therefore, only if the rate of growth influences the effort-supply decision, are any effects on wages and unemployment to be expected. In line with the other chapters, the effort function is based on the fair-wage efficiency theory, but, as will be shown below in more detail, the results do not depend on which version of efficiency-wage model is chosen.

5.2.2.1 The Model

For the same reasons as in the previous section, with one aggregate sector, effort supply is now not only a function of the wage that the own firm pays, but also positively related to the unemployment rate and negatively to the level of unemployment benefits. High unemployment rates and hence a relatively long expected unemployment duration leads to a relatively high expected income loss and hence individuals will increase their effort supply in order to forego a potential dismissal.[5] An opposite line of reasoning holds for high unemployment benefits. Therefore, the effort function can be written as

$$\tilde{e}_i = \tilde{e}\left(\frac{\hat{w}_i}{\hat{w}^e}, u, \hat{z}\right) \tag{5.20}$$

with $\partial\tilde{e}/\partial(\hat{w}_i/\hat{w}^e) > 0$, $\partial\tilde{e}/\partial u > 0$ and $\partial\tilde{e}/\partial\hat{z} < 0$ where $\hat{w}_i$ is the wage paid by firm i and $\hat{w}^e$ is the wage a worker expects to obtain outside of the firm.[6] As both the unemployment rate and the real level of unemployment benefits $\hat{z}$ are treated as exogenous by the firm, the wage that the firm pays is the only variable it can set to influence the effort that workers exert. Further, assuming that workers directly compare their own wage with the expected outside wage means that the effort function is homogeneous of degree zero with respect to the absolute wage levels. Therefore, the Solow condition as originally given by equation (3.17), holds in a slightly modified form as

$$\frac{\partial\tilde{e}\left(\frac{\hat{w}_i}{\hat{w}^e}, u, \hat{z}\right)}{\partial\frac{\hat{w}_i}{\hat{w}^e}} \frac{\frac{\hat{w}_i}{\hat{w}^e}}{\tilde{e}\left(\frac{\hat{w}_i}{\hat{w}^e}, u, \hat{z}\right)} = 1 \tag{5.21}$$

---

[5] See, for example, Kahneman et al. (1986) for empirical evidence.
[6] For example, modifying the effort function (4.37) to

$$\tilde{e}_i = -\varepsilon_1 + \varepsilon_2 \left(\frac{\hat{w}_i}{\hat{w}^e}\right)^{\frac{u\varepsilon_3}{\hat{z}}}$$

fulfills these properties.

In the steady-state general equilibrium with $\hat{w}_i = \hat{w}^e = \hat{w}^*$, this equation uniquely defines the unemployment rate solely as a function of the effort-function parameters and the level of unemployment benefits. Thus, with firms hiring labour and setting the wage level to minimise unit costs which only depend on the parameters of the effort function, the steady-state unemployment level is independent of the growth rate. In other words, as the growth rate does not influence the effort-supply decision, it is also not important for the wage setting process and hence has no employment effects.

5.2.2.2 Discussion

As shown above, the rate of technological progress does not influence the effort function. Further, assuming as is theoretically and empirically plausible, that the effort function is homogeneous of degree zero with respect to the wage level paid in the firm under consideration relative to the wage paid by other firms, means that the Solow condition uniquely defines the unemployment rate only as a function of the effort-function parameters. In addition, seeing as, with the exception of the parameters of the effort function, there is no mechanism which prevents firms from lowering wages, any change in the growth rate which leads to a change in the optimal capital intensity as given by equation (5.11), simply means that the wage rate in all firms will adapt. Therefore, in the steady-state equilibrium, the growth rate has no influence on the unemployment rate.

The results derived so far in this chapter have made it clear, that the effects of growth on the unemployment rate crucially depend on the cause of the labour-market imperfection. For this reason and in line with the other chapters, the next section analyses this relationship when the labour market is characterised by matching frictions.

5.2.3 Matching Processes

Although the matching models analysed in Chapters 3.3 and 4.2.3 were dynamic in the sense that future utility and expected profits determined labour supply and demand behaviour respectively, the models in these chapters firstly assumed that the firm only used labour as variable input with capital fixed and secondly, that the production technology and hence productivity of a worker were constant. However, it is to be intuitively expected that the introduction of capital and technological progress in the production function affect the returns that a firm can expect from opening up a new position. Thus, these variables influence the job-creation condition for firms and will therefore affect the unemployment rate. These ideas were first formalised by Eriksson (1997) which builds the basis for the following section.

5.2.3.1 The Model

In the basic-matching model analysed in Chapter 3.3, the productivity of a worker was assumed to be constant. Here, with labour-augmenting technological progress, productivity is now a function of time with $y(0) = A(0)$ and hence $y(t)$ grows at the rate g. Further, as technological change is assumed to be disembodied, this rate of productivity growth affects existing and new matches alike.

As outlined above, production now also requires capital as an input factor. Therefore, omitting a firm subscript for notational simplicity, the present-discounted value of expected profits is

$$\pi = \int_0^\infty e^{-rt} \left[\hat{F}(\hat{k}) - wL - cA\Upsilon - I \right] dt \tag{5.22}$$

where Υ denotes the number of vacancies within a firm. Further, with $f(\theta)$ as the rate at which vacancies are filled and b the rate at which jobs terminate, the size of the labour force within a firm fluctuates according to

$$\dot{L} = f(\theta)\Upsilon - bL \tag{5.23}$$

Therefore, dynamic optimisation with the number of vacancies Υ and investments I as the control and the size of the labour force and capital stock as the state variables, and making use of (5.5) which defines labour's marginal productivity leads to

$$\hat{w} = \hat{F}(\hat{k}) - \hat{k}\hat{F}'(\hat{k}) - \frac{b + r - g}{f(\theta)} cA \tag{5.24}$$

i.e. the wage is a function of the marginal productivity of labour and the expected capitalised value of the firm's hiring costs.

Assuming as throughout that there is a perfectly competitive capital market, means that firms will always be able to obtain capital at the going interest rate. For this reason and because renting capital is costly, even if a firm posts a new vacancy, it will not rent the required capital stock until a suitable worker has been found. Therefore, with the exception of the increasing costs of posting a vacancy, the present-discounted value of a vacant position remains unchanged and is given by

$$rW^V = f(\theta)(W^F - W^V) - cA \tag{5.25}$$

However, once the position is filled, the value of the capital stock that is required needs to be taken into account so that the present-discounted value is now given by $W^F + A\hat{k}$. Further, due to the fact that capital depreciates, the net return to filling a position also needs to be modified with the Bellman equation for the present-discounted value of a filled position now

$$r(W^F + A\hat{k}) = A\hat{F}(\hat{k}) - \delta A\hat{k} - \hat{w} - bW^F \qquad (5.26)$$

As the firm can freely buy and sell capital, the firm faces no commitment with respect to its capital stock as it is liquid and reversible. Therefore, the worker cannot hold up the firm and the Nash wage bargain between firms and employees is unaffected by the introduction of capital. Hence, workers still receive a share $\tilde{\beta}$ of the surplus. Similarly, the present-discounted value of lifetime utility for an employed individual, given by equation (3.34), also holds so that these equation are simply repeated here for convenience

$$\tilde{V}^E - \tilde{V}^U = \tilde{\beta}(W^F - W^V + \tilde{V}^E - \tilde{V}^U) \qquad (5.27)$$
$$r\tilde{V}^E = \hat{w} - b(\tilde{V}^E - \tilde{V}^U) \qquad (5.28)$$

This last equation together with the modified Bellman equation for a filled position as given by (5.26) and the fact that in a steady state all job opportunities are exploited so that $W^V = 0$ holds, can be inserted into equation (5.27) to yield

$$\frac{\hat{w} + b\tilde{V}^U}{r + b} - \tilde{V}^U = \tilde{\beta}\left(\frac{A\hat{F}(\hat{k}) - \delta A\hat{k} - rA\hat{K} - \hat{w}}{r + b} + \frac{\hat{w} + b\tilde{V}^U}{r + b} - \tilde{V}^U\right)$$

from which the wage rate is derived as

$$\hat{w} = \tilde{\beta}(A\hat{F}(\hat{k}) - (r + \delta)A\hat{k}) + (1 - \tilde{\beta})r\tilde{V}^U \qquad (5.29)$$

With the present-discounted value of currently being unemployed also dependent on the technological level as these determine the returns to search, this value is given by

$$r\tilde{V}^U = z + \frac{\tilde{\beta}}{1 - \tilde{\beta}}cA\theta \qquad (5.30)$$

Inserting this equation and (5.8) into (5.29) gives

$$\hat{w} = \tilde{\beta}(\hat{F}(\hat{k}) - \hat{k}\hat{F}'(\hat{k}) + c\theta) + (1 - \tilde{\beta})\hat{z} \qquad (5.31)$$

In a final step, using the intertemporal optimisation results (5.8) and (5.12) whereby

$$r = \rho + \gamma g \qquad (5.32)$$

the wage equations (5.24) and (5.31) can be combined to yield the modified job-creation condition

$$\left(1 - \tilde{\beta}\right)\left(\hat{F}(\hat{k}) - \hat{k}\hat{F}'(\hat{k})\right) - \frac{b + \rho - g(1 - \gamma)}{f(\theta)}c - \left(1 - \tilde{\beta}\right)\hat{z} - \tilde{\beta}c\theta = 0 \quad (5.33)$$

Applying the implicit-function theorem, this condition can be used to derive the effects of a higher growth rate on job creation as

$$\frac{d\theta}{dg} = -\frac{(1 - \tilde{\beta})\hat{k}\gamma - \frac{1-\gamma}{f(\theta)}c}{\tilde{\beta}c - \frac{\rho+b-g(1-\gamma)}{\theta f(\theta)}c\epsilon_{f,\theta}} \quad (5.34)$$

where $\epsilon_{f,\theta}$ is the elasticity of the job-finding rate with respect to labour-market tightness which by the assumptions of the properties of the matching function takes on a value in the range $(-1,0)$. By assumption $\rho > g$ holds so that the denominator is unambiguously positive.[7] For $\gamma \geq 1$, the numerator is also positive, and hence in this case there is a negative relationship between the growth rate and labour-market tightness.[8] As the introduction of technological progress does not alter the steady-state flow condition for the labour market so that the Beveridge curve

$$u = \frac{b}{b + f(\theta)\theta} \quad (5.35)$$

continues to hold, if $\gamma \geq 1$, a higher growth rate leads to a higher unemployment rate. However, these results may be overturned for sufficiently low values of γ. The intuition for this result can be best understood from (5.32). As outlined above, a higher growth rate increases the effective depreciation rate of capital so that capital's marginal productivity and hence the interest rate must rise accordingly. How high this rise in interest rate needs to be can be seen from the above equation and crucially depends on the value of the elasticity of marginal utility γ. Thus, with "high" values of this parameter (and correspondingly small values of the intertemporal elasticity of substitution), consumers have a relatively low inclination to save and increase their future consumption level so that a large increase in the interest rate is necessary to induce consumers to save more. However, the resulting higher rate at which future earnings are discounted means that the present-discounted value of profits from a vacancy decrease so that fewer positions are created and the unemployment rate will rise. However, a higher growth rate also lowers $r - g$, the effective discount rate of firms, because a higher rate of technological progress means that future output and hence profits also grow at a faster rate, inducing firms to increase their job creation. As can be seen from (5.32), for values of $\gamma < 1$, the effective discount rate will fall. In addition, a

---

[7] If $\rho < g$ were assumed, then future real consumption would be valued more than contemporary consumption and the utility problem would no longer be bounded.

[8] Substantial empirical work has been devoted to estimating the intertemporal elasticity of substitution $1/\gamma$. The general consensus suggests a value below unity. See, for example Blundell (1991) and more recently Berloffa (1997) and the works cited therein.

higher growth rate means that the costs of posting a vacancy in the future increase at a faster rate. Therefore, since the costs of a vacancy are borne today but profits from the match accrue in the future, a firm can save costs by increasing current job creation leading to a fall in the unemployment rate.[9] Hence, which effect dominates is an empirical question but with the bulk of the evidence indicating a low value for the intertemporal elasticity of substitution and hence a high value for γ, unemployment is likely to increase with higher growth as firms reduce the amount of job creation.

5.2.3.2 Discussion

As shown above, when labour markets are characterised by matching frictions, whether the unemployment rate rises or falls with higher growth rates cannot be unambiguously determined. On the one hand, a higher growth rate increases the rate at which the costs of posting a vacancy rise in the future. This will make it more attractive for firms to pull forward planned job openings. On the other hand, a higher interest rate also influences the "effective" discount rate $r - g$. Whether this rate increases or decreases crucially depends on the intertemporal substitution elasticity as this determines the extent to which the interest rate increases at higher growth levels. For high levels of this elasticity, consumers are relatively willing to have different consumption levels over time and hence, the new equilibrium is characterised by only a small increase in the interest rate so that the effective discount rate decreases. This will increase the value of a position and further induce firms to post additional vacancies. Therefore, in this case, higher growth rates are associated with lower equilibrium unemployment rates. These results, however, are overturned for (empirically more plausible) low values of the intertemporal substitution elasticity in which case the rise in the interest rate causes net returns from posting a vacancy to fall so that fewer jobs will now be created and the unemployment rate increases.

5.3 Summary and Conclusions

As stated at the beginning of the chapter, there are numerous effects of higher growth on unemployment and hence a certain degree of ambiguity as to whether a higher growth rate is harmful for employment or not. Unfortunately, this *ex-ante* ambiguity still exists even after a formal analysis, with the effects depending on the type of labour-market imperfection. Thus,

---

[9] This effect has been termed the "capitalisation effect" by Aghion and Howitt (1994).

when the labour market is characterised by the presence of unions bargaining over wages with employers, a higher growth rate leading to higher labour productivity and hence lower wage costs, will unambiguously lead to a reduction in the unemployment rate. However, this result no longer holds when wages are set by efficiency-wage considerations or when the labour market is characterised by matching frictions. In the case of efficiency wages, firms can unilaterally set both the wage and employment level in order to minimise their unit-labour costs. Thus, the profit-maximising wage (and employment) level is determined by the parameters of the effort function. However, even if in the economy as a whole the amount of effort individuals exert is positively related to the unemployment rate, a single firm is too small to influence the unemployment rate and will therefore treat this variable as exogenous. Thus, although a change in the growth rate of the economy will lead to a change in the wage rate, this change will influence all firms identically so that the relative wage and hence effort supply remains constant. Therefore, any change in the economy's growth rate and labour productivity will be completely absorbed by a change in the wage rate and therefore leaves the (un)employment rate unaltered. In the case of matching frictions, the result crucially depend on the size of the household intertemporal substitution elasticity as this determines whether the present-discounted value of a position rises or falls at higher growth rates and hence whether it becomes more or less attractive to open up new positions. For empirically plausible values of this elasticity, a higher growth rate will lead to a fairly strong increase in the interest rate causing the return from posting a vacancy to fall. Therefore, although job destruction is left unaltered, the degree of job creation decreases so that the unemployment rate increases.

Summing up, the above analysis shows that unemployment can rise, fall or remain constant with higher growth rates depending on which model of the labour market and which values for certain variables are assumed. This makes clear that further analysis is required before clear conclusions can be made whether growth is "good" or "bad" for unemployment.

One obvious shortcoming of the models in this chapter is the fact that the source of growth was assumed to be exogenous. Whilst this is understandable if one accepts that the early growth economists thought that the source of growth was of technical nature and could hence be better explained by the engineering sciences, this assumption is clearly unsatisfactory in the sense that there are no policy recommendations from the above model with which the growth rate could be influenced. In addition, with the growth rate exogenously given, it is not possible to analyse whether the presence of, for example, unions in the labour market raises or lowers the growth rate. Further, as shown in Chapter 4, the degree of product-market competition also plays a decisive role in determining the unemployment rate. It is for these reasons that in the next chapter, product markets are assumed to be imperfectly competitive and the source of growth is endogenised.

6

Market Structure, Innovation-Based Growth and Unemployment

The previous chapter has shown that there is an ambiguous relationship between growth and unemployment. However, these results were all derived under the assumption that growth resulted from exogenously given technological progress. As these neoclassical growth models were able to explain many empirical findings such as decreasing growth rates and conditional convergence well, research neglected the issue of a growth for a while. However, by the mid 1980s with the gap between the developed and developing world continuing to increase, but also in order to be able to predict how economies could increase their growth rates following the two oil crisis, it became clear that the mechanisms driving growth needed to be described more precisely and that new theories which endogenously explained growth were needed.[1] As can easily be shown, any factor leading even to a small permanent percentage increase in the annual growth rate has substantial consequences for the long-term per-capita income level. Thus, as Lucas (1988, p. 5) puts it:

> *"Once one starts to think about them* [growth rates]*, it is hard to think about anything else"*

In contrast to the neoclassical growth theory which is based on one model with only minor deviations, modern growth theory is characterised by a multitude of different mechanisms, each giving an alternative explanation for endogenous growth. The only common feature that these models have is that they are no longer characterised by diminishing returns. For example, in the model by Romer (1986), investment into physical capital leads to new technological knowledge as a by-product. Even if this new technological knowledge is developed by private firms, the resulting blueprints are a public good, i.e. there is unlimited access to this new knowledge and it can be used simultaneously by more than one firm as a building block towards developing further technological improvements. Thus, as opposed to physical capital in the neoclassical

---

[1] For a detailed overview of these endogenous growth theories, see Barro and Sala-i-Martin (1995).

growth model, even as the economy accumulates more knowledge, the gains from further knowledge do not diminish. Similarly, in the model by Lucas (1988), it is human capital which drives economic growth. This is general knowledge obtained during schooling which is passed on from one generation to the next. Each generation uses this obtained knowledge and through own research, enlarges this knowledge base further. This higher knowledge level directly increases worker productivity, thereby leading to positive, permanent growth.

Nevertheless, that technological progress is one of the primary factors determining the growth rate of an economy as assumed in the neoclassical growth model is without doubt. However, both the models by Romer (1986) and Lucas (1988) cannot explain how, and perhaps more importantly why, technological change occurs. For this reason, a whole new class of endogenous growth models emerged which have R&D leading to new innovations as the central driving force of the growth rate of an economy. The two seminal models of this class are by Grossman and Helpman (1991a,b), where innovations can either lead to an increasing quantity of goods (so-called "horizontal" innovations), or to a rising quality of the set of available goods ("vertical" innovations), and by Aghion and Howitt (1992, 1994) who only model vertical innovations.[2] In the case of a constantly increasing number of products, either total productivity of firms increases as they can use a greater variety of intermediates as inputs in their production, or consumers are assumed to have a "love of variety" and have a constantly rising number of varieties to choose from. Thus, the theory of monopolistic competition as described in Chapter 4, became an integral part of growth theory and marks the departure of growth theorists from the assumption that the economy is characterised by perfectly competitive markets. In the case of horizontal innovations, the price charged cannot be "too high", as otherwise firms or consumers will prefer to switch demand to a different variant which leads to decreasing profits. If innovations are of a vertical nature, then it is assumed that these innovations are protected from imitation by a perfectly enforceable infinite patent. However, a firm is still restricted in the price it can demand as consumers will only buy the product if the quality-adjusted price is (marginally) below that of the next lower quality product. Therefore, irrespective of the nature of the technological change, firms earn (at least temporary) positive profits. It is the lure of these profits that is the driving force behind continuous investments into R&D and thus long-run growth is measured either as a constantly increasing number of products, i.e. quantitative growth, or in an increasing quality of the products, i.e. qualitative growth.[3]

---

[2] See, Aghion and Howitt (1998) for an overview of R&D-based growth models.

[3] The underlying idea of profits being the driving force behind innovations is far older than these models and goes back to Schumpeter (1942). For this reason, models of this type are also called Schumpeterian growth models.

As Aghion and Howitt (1992, 1994) can be treated as a special case (with only one intermediate good which is improved stochastically) of the Grossman and Helpman (1991a,b) models, it is this latter class that builds the basis for the analysis in this chapter. Further, as within these models the qualitative results are almost identical irrespective of whether innovations are assumed to increase the quantity or quality of goods, but from an empirical viewpoint it seems to be primarily the quality of computers, cars etc. that has increased, it is this type of growth that is analysed here.[4]

As will be shown below in more detail, the probability that research activities succeed, i.e. lead to the invention of a qualitatively higher good, is random and governed by a homogeneous Poisson process whose intensity, i.e. the expected number of innovations per unit time, equals the effective investment into R&D. Further, if it is assumed that each innovation improves the quality of the good by a constant factor λ, and the number of past improvements is given by $\bar{m}$, then over time, if the initial quality at t_0 is normalised to 1, the quality of the product can be depicted as a "quality ladder" as shown in Fig. 6.1 where each innovation and rung up the ladder marks the introduction of a new "state-of-the-art" technology. Thus, although in reality goods are judged by many

Fig. 6.1: Stochastic Intervals Between Quality Improvements along a "Quality Ladder"

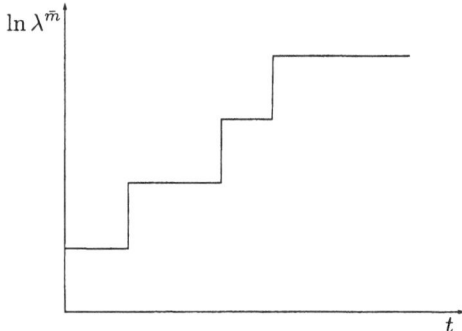

characteristics so that it is not always obvious which good is "better", here quality is assumed to be objectively measurable and the only attribute by which the goods differ.

As in Chapter 4 where the composite manufacturing good M was comprised of n different brands, here this composite good consists of a bundle of intermediate goods j. If research is aimed at all intermediate goods, and the improvements of each intermediate good is governed by a homogeneous but independent Poisson process, then because the realisation of research success

---

[4] See, for example, Dinopoulos and Thompson (1998), Young (1998) or Li (2000) for models with both horizontal and vertical innovations, however, all in the context of perfectly competitive labour markets.

is stochastic, over time the number of quality improvements will be different for each intermediate good as shown in Fig. 6.2. However, if the number of

Fig. 6.2: Number of Improvements of each Intermediate Good

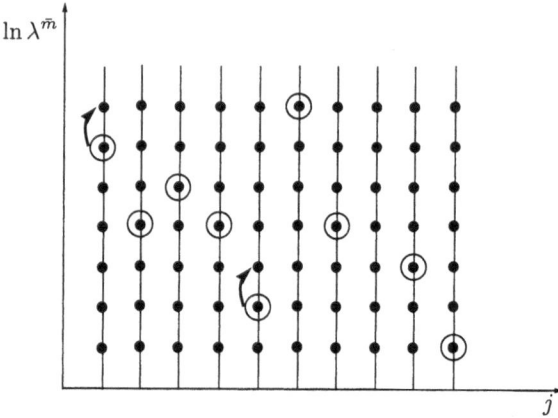

different intermediate goods is "large" (meaning that the *law of large numbers* holds), then the Poisson parameter of research success also defines the expected number of innovations across all intermediate goods per unit time.

With such an endogenous engine of growth, it is possible to not only analyse how unemployment and growth are related, but also how different wage-setting mechanisms influence the growth rate itself. In addition, as shown in Chapter 2, the labour-market performance of high and low-skilled workers varies dramatically. In the United States the wage differential between these two types of labour has risen substantially, whereas with the exception of the United Kingdom, in western Europe the low-skilled have suffered far more from increasing unemployment rates than falling relative wages. The worsening position of the low-skilled in Germany can clearly be seen by the "Greencard" debate in Germany by which firms wanted a large number of software programmers to be enticed to come and work in Germany for several years. This highlights the fact that there are some branches of the economy which are not willing to hire the unemployed but seem to be suffering from a severe shortage of workers of the appropriate skill level.[5] In order to incorporate this fact, the models in this chapter will treat labour as heterogeneous with respect to the skill level. However, before turning to the labour market, the other sectors of the economy will be analysed more formally as these are independent of the wage-setting mechanism. Subsequently, as throughout, the analysis will

---

[5] There are several factors why only few foreigners made use of this offer. Firstly, the offer had severe restrictions regarding how long and which family members were allowed to come. Secondly, German is not a langauge spoken by many foreigners and thirdly, the recent worsening of the economic situation in Germany has reduced the demand for IT-workers.

first be exemplified under the assumption that wages are set by negotiations between unions and employers, then by efficiency wages and finally, the effects on growth and unemployment in the presence of matching frictions will be studied.

6.1 Household and Producer Decisions

6.1.1 Households

As in the previous chapters, households are modelled as infinitely-lived dynasties. These maximise their intertemporal utility gained from the consumption of two different goods: a traditional good and a high-tech manufacturing good. At any point in time, each household is comprised of a fraction ϕ workers who are high-skilled and $(1 - \phi)$ who are low-skilled workers. Further, due to labour-market imperfections, some of these household members will be currently unemployed. Therefore, the dynamic optimisation decision faced by households is very similar to that described in Chapter 4.1.1 and can be broken up into two stages. In the first stage, households must decide how to divide their income between consumption and savings subject to their budget constraint. However, as will be shown below in more detail, due to qualitative growth, the relative price of the manufacturing good is not constant. For this reason, the result of this first stage leads to the Keynes–Ramsey rule derived in Chapter 4.1.1 and given by

$$\frac{\dot{C}}{C} = \frac{1}{\gamma}\left(r - \frac{\dot{P}}{P} - \rho\right) \tag{6.1}$$

to take the effects of changes in the macroeconomic price index P into account.

In the second stage, the consumers must optimally determine the share of expenditures devoted to each of the goods. Assuming the same consumption function as in Chapter 4, i.e.

$$C = M^{\varsigma}T^{1-\varsigma} \tag{6.2}$$

again leads to these shares being given by

$$p_M M = \varsigma PC \tag{6.3}$$

for the manufacturing good and

$$p_T T = (1 - \varsigma)PC \tag{6.4}$$

as the expenditure share spent on traditional goods.

As mentioned above, the price index is now no longer constant due to repeated quality improvements of the manufacturing good. In order to be able
to analyse this process more thoroughly, it is necessary to take a closer look
at the production side of the economy.

6.1.2 Firms

There are two final goods produced in the economy: a "traditional" homogeneous good which is not improved over time and a manufacturing good whose
quality increases with time. Turning first to the traditional sector, then the
production technology in this sector is assumed to be given by a CES production function which has low- and high-skilled labour as imperfect substitutes
for each other but with an exogenously fixed elasticity of substitution σ_T
between them. Further, the output elasticities of the two types of labour are
assumed to be constant and sum to one so that the technology exhibits constant returns to scale. However, it is assumed that this sector has the lowest
high-skilled labour intensity. Finally, firms operating in this sector face perfect
competition on the goods market. Therefore, the price p_T of the traditional
good is given by

$$p_T = c_T(w_{LT}, w_{HT}) \tag{6.5}$$

where w_{LT} and w_{HT} are the (nominal) wage rates of low- and high-skilled
labour respectively, and c_T denotes unit costs in this sector. The resulting
demand for low-skilled labour can be derived by applying Shephard's lemma
which yields

$$L_{LT} = a_{LT}(w_{LT}, w_{HT})T \tag{6.6}$$

with a_{LT} as the variable input coefficients which are derived from the partial
derivative of the unit cost function with respect to low-skilled wages. The
corresponding demand for high-skilled labour can be derived analogously as

$$L_{HT} = a_{HT}(w_{LT}, w_{HT})T \tag{6.7}$$

The other final good is a high-technology manufacturing good which is produced using a whole range of intermediates j. As growth in this model is of
qualitative nature, the set of available intermediate goods is constant. Due to
research activities, the quality of these intermediate goods is constantly being
improved. Denoting the quality of the $\bar{m}$th generation of intermediate good j
by $q_{\bar{m}_j}$, and assuming a constant returns to scale Cobb–Douglas technology
means that output of this composite good M is given by the input index

$$M = \left(\sum_{\bar{m}} q_{\bar{m}_1} x_1 \right)^{\frac{1}{\bar{n}}} \left(\sum_{\bar{m}} q_{\bar{m}_2} x_2 \right)^{\frac{1}{\bar{n}}} \cdots \left(\sum_{\bar{m}} q_{\bar{m}_{\bar{n}}} x_{\bar{n}} \right)^{\frac{1}{\bar{n}}}$$

where $x_j, j = 1, \ldots, \bar{n}$ denotes the quantities of the $\bar{n}$ different input goods. This index can be reformulated as

$$\ln M = \ln \left[\sum_{\bar{m}} q_{\bar{m}_1} x_1 \right]^{\frac{1}{\bar{n}}} + \ln \left[\sum_{\bar{m}} q_{\bar{m}_2} x_2 \right]^{\frac{1}{\bar{n}}} + \cdots + \ln \left[\sum_{\bar{m}} q_{\bar{m}_{\bar{n}}} x_{\bar{n}} \right]^{\frac{1}{\bar{n}}}$$

In order to simplify the analysis, the set of intermediate goods j is normalised to vary continuously in the unit interval. With this assumption, the index becomes

$$\ln M = \int_0^1 \ln \left[\sum_{\bar{m}} q_{\bar{m}_j} x_j \right] \mathrm{d}j \tag{6.8}$$

Every time there is a successful innovation targeted at the intermediate j, its quality is improved by a constant factor $\lambda > 1$.[6] Thus, $q_{\bar{m}_j} = \lambda q_{\bar{m}_j - 1}$ holds for all $\bar{m}$. Therefore, normalising the quality of each variant at time t_0 to one, means that the present quality of an intermediate good j is given by $\lambda^{\bar{m}_j}$ which defines the "state-of-the-art". As shown below, firms producing the current highest quality good will charge a price so that their good just has the lowest quality-adjusted price relative to its predecessor. This means that only the highest quality good in each intermediate sector will be demanded. Hence, the above composite good index M simplifies to

$$\ln M = \int_0^1 \ln \left[\lambda^{\bar{m}_j} x_j \right] \mathrm{d}j \tag{6.9}$$

As the qualities of the components in each intermediate market j differ, but any two variants within one market are identical in all other respects and hence perfect substitutes for each other (as long as the appropriate adjustment for quality differences is made), the technological leader has the ability to capture the entire market demand by charging a quality-adjusted price which is marginally lower than that of his or her nearest competitor. Denoting the price of intermediate good j by p_j, then the minimum quality-adjusted of this good is $p_j / \lambda^{\bar{m}_j}$. Therefore, from (6.9), the price index p_M is determined as

$$p_M = \exp \left\{ \int_0^1 \ln \left[\frac{p_j}{\lambda^{\bar{m}_j}} \right] \mathrm{d}j \right\} \tag{6.10}$$

It can also be seen from equation (6.9) that the intermediate goods in each market j are perfect substitutes for another if one adjusts for quality. Further, seeing as the elasticity of substitution between any pair of intermediate

---

[6] See Stadler (2001) for a model in which the size of the quality improvement can vary amongst the different intermediates and Li (2002) where quality progress is modelled as a continuous process rather than a series of discrete jumps.

products is equal to one, all components will be used in equal quantities. Therefore, the manufacturing-good index becomes

$$M = A_M X \qquad (6.11)$$

where A_M defines the average quality and $X = x$ the aggregate output (as the number of varieties has been normalised to unity) of intermediates. If each industry has its own Poisson process with instantaneous arrival rate ι_j, summing over all industries means that in the time interval τ, the total number of expected quality improvements will be $\tilde{\iota}(\tau) = \int_0^\tau \iota(t)\, dt$. This means that the average quality of the varieties is

$$A_M(\tau) = \lambda^{\tilde{\iota}(\tau)} \qquad (6.12)$$

Production of the intermediate goods is assumed to be more human capital intensive than in the traditional sector, but less intensive than in the research sector. All firms producing the intermediate goods $j \in [0,1]$ have identical unit-cost functions given by

$$c_j(w_{L_j}, w_{H_j}) \qquad (6.13)$$

where w_{L_j} and w_{H_j} are the wage rates paid by the firm producing the intermediate good j for the low- and high-skilled workers respectively. From this cost function, the low-skilled labour demand in the intermediate goods sector is derived as

$$L_{LX} = a_{LX}(w_{LX}, w_{HX})X \qquad (6.14)$$

and

$$L_{HX} = a_{HX}(w_{LX}, w_{HX})X \qquad (6.15)$$

for high-skilled labour, where a_{sX} denotes the partial derivative of the cost function in the intermediate goods sector with respect to the wage rate for labour of skill-group $s \in \{L, H\}$. Further, from the cost equation (6.13), the optimal "limit pricing" strategy is[7]

---

[7] In the specification of the model as above, the elasticity of substitution between any two components is assumed to be unity. Relaxing this assumption means reformulating the manufacturing good index (6.9) as

$$\ln M = \left\{ \int_0^1 \ln \left[\lambda^{\tilde{m}_j} x_j \right]^\kappa \, dj \right\}^{\frac{1}{\kappa}}$$

so that $\sigma \equiv 1/(1-\kappa)$ is the elasticity of substitution between any two variants. In this case, profit-maximisation leads to a markup factor of $1/\kappa$. For the case that innovations are "large" in the sense that $\lambda > 1/\kappa$, then intermediate-good pro-

$$p_j = \lambda c_j(w_{L_j}, w_{H_j}) \tag{6.16}$$

which leads to profits π_j for a market leader for the intermediate good j of

$$\begin{aligned} \pi_j &= p_j x_j - c_j x_j \\ &= p_j x_j (1 - 1/\lambda) \end{aligned} \tag{6.17}$$

If market leaders invest in research efforts aimed at the further improvement of their good, then by the same argument as above, the maximum price they could charge would be $\lambda^2 c_j$. In this case they would earn a profit of $p_j x_j (1 - 1/\lambda^2)$. This means that the marginal gain from being two steps ahead of the closest rival is $(1/\lambda)(1 - 1/\lambda)$ which is strictly less than the additional profit that results if it is possible to displace a current monopolist in another sector. Therefore, due to this "replacement effect" (Tirole 1988, p. 392), it is never optimal for the current market leader in sector j to undertake further research aimed at improving the quality of his own product. The knowledge required to improve the quality of the intermediate goods is obtained by researchers working in a separate sector discussed in the next section.

6.2 Research Decisions

The research technology also exhibits constant returns to scale and requires both low- and high-skilled labour. The research sector is assumed to have the highest ratio of high- to low-skilled workers. The unit cost function in this sector is given by $c_R(w_{LR}, w_{HR})$, with w_{sR} the wage rate in the research sector for labour of type $s \in \{L, H\}$. From this it is possible to derive the labour demand needed to achieve a research intensity ι as

$$L_{LR} = a_{LR}(w_{LR}, w_{HR})\iota \tag{6.18}$$
$$L_{HR} = a_{HR}(w_{LR}, w_{HR})\iota \tag{6.19}$$

with a_{sR} as the respective partial derivatives of the cost function with respect to low- and high-skilled labour.

Innovations are financed through the emission of stocks. If a firm has a successful innovation, it will achieve a stock market value of $\bar{V}$ arising from the expected-profit streams that a market leader enjoys. Therefore, any investor

---

ducers pricing decisions are unconstrained by potential competition from owners of previous patents. Aghion and Howitt (1998, Chap. 2) label these innovations "drastic". For all innovation sizes smaller than $1/\kappa$, the price charged by a current monopolist is limited by λ. Thus, although non-drastic innovations are assumed in the following, all results could easily be generalised to the case of drastic innovations.

can purchase a share which, during the time interval dt, pays a dividend of $\bar{V}$ dt with probability ι dt. At a research intensity of ι, total research costs during the time interval dt are given by $(a_{LR}w_{LR} + a_{HR}w_{HR})\iota$ dt. Given unit research costs of c_R, each research firm maximises $\bar{V}\iota$ d$t - c_R\iota$ dt which requires an infinite amount of research if $\bar{V} > c_R$ and zero research and no quality improvements if $\bar{V} < c_R$. Hence, finite research investments which lead to positive quality growth in equilibrium only occur if

$$\bar{V} = c_R(w_{LR}, w_{HR}) \tag{6.20}$$

Whether these research investments lead to a successful innovation or not is governed by a Poisson process. As each innovation leads to a higher quality intermediate good used in the production of the manufacturing good, the rate at which the quality of this composite good grows is given by[8]

$$\frac{\dot{M}}{M} = \iota \ln \lambda \tag{6.21}$$

Thus, the growth rate of the quality of the composite good depends on the size of innovations weighted with the probability that a firm successfully innovates.

The analysis so far resembles the original Grossman and Helpman (1991a, Chap. 5) model. However, they assume labour markets that always clear. This assumption is dropped in the next sections where the influence of different labour-market imperfections are introduced.

6.3 Wage Setting and General Equilibrium

6.3.1 Union Wage Bargaining

Whereas in Chapter 5.2.1 the effects of growth on unemployment when wages are set by unions was analysed, this chapter is also concerned with the question of how unions effect the growth rate itself.[9] As shown in Chapter 2, in the past two decades continental Europe has witnessed relatively low growth and high unemployment rates whereas the US had sustained high growth rates and more or less full employment levels. One of the reasons often stated for this divergence in economic performance is, as Nickell and Layard (1999, p. 3030) put it,

---

[8] See Appendix A.7 for the derivation.

[9] See Nickell and Layard (1999) for an analysis of how other labour market institutional settings such as taxation, unemployment benefits, minimum wages, active labour-market policies and the general education system affect the growth rate.

"Barely a day goes by without some expert telling us how the continental European economies are about to disintegrate unless their labor markets become more flexible. Basically, we are told, Europe has the wrong sort of labor market institutions for the modern global economy. These outdated institutions both raise unemployment and lower growth rates."

One of the primary candidates for such an "outdated institution" often mentioned in this context is the presence of strong labour unions which hinder structural change towards expanding industries – see, for example, Eichengreen and Iversen (1999) for this view. To analyse this viewpoint more formally, the model of labour unions with heterogeneous workers as presented in Chapter 3.1.2 is integrated into the R&D-based growth model outlined above.

6.3.1.1 The Model

One of the main justifications for unions and one of the reasons why workers decide to become members is the expected redistribution effect. In the present model, this means that on the one hand it is the low-skilled workers who can profit most from union membership, with the aim of reducing the wage differential between the two skill groups. On the other hand, this implies that high-skilled workers have no incentive to join the union and the labour market for these workers is fully competitive.[10] Although trade unions do not publish official figures on the skill level of their members, the assumption that it is primarily the low-skilled who are organised in trade unions is justified from an empirical viewpoint. For example, in Germany of all the members in the trade union representing workers in the metal industry, by far the largest union in the German trade union federation, over 83% were "blue collar" workers – see Deutscher Gewerkschaftsbund (2002) for more details. Further, in the German Socio-Economic Panel data of 1998, fewer than 10% of all union members had a tertiary education degree.[11]

As shown above, firms producing the highest quality good will demand a limit price higher than their marginal costs and therefore earn positive profits as long as they possess the leading technology. Therefore, there are only rents in the intermediate sector and unions only have an incentive to operate here. Therefore, assuming a "right-to-manage" approach as throughout, the Nash product is

---

[10] See Palokangas (1996) for a model in which unions represent both types of labour simultaneously. However, here, as long as the economy exhibits a positive growth rate, the high-skilled are fully employed.

[11] I thank Harald Strotmann from the IAW Tübingen, for performing the necessary calculations.

$$\Omega = \max_{w_{L_j}} \left[(w_{L_j} - \bar{w}_{LX}) L_{Lj} \right]^{\beta_L} [\pi_j - \pi_0]^{1-\beta_L} \qquad (6.22)$$

where L_{Lj} is the firm's low-skilled labour demand and $\bar{w}_{LX}$ the low-skilled reservation wage in the intermediate sector. Assuming that the two types of labour are gross complements as inputs in production, means that if no wage agreement is reached with the union and the low-skilled workers go on strike, the firm is forced to completely stop production and therefore has a negative fallback position. Thus, maximisation of the above equation leads to a low-skilled nominal wage given by

$$w_{L_j} = \left[1 - \frac{\beta_L(1 - \epsilon_{F,L_L})}{\beta_L(1 - \epsilon_{F,L_L})\epsilon_{L_j,w_L} - \epsilon_{F,L_L} + \beta_L + \omega(1 - \beta_L)} \right] \bar{w}_{LX} \qquad (6.23)$$

where $\epsilon_{L_j,w_L} < 0$ denotes the elasticity of low-skilled labour demand with respect to low-skilled wages in the intermediate sector and ϵ_{F,L_L} is the output elasticity with respect to low-skilled labour, both of which will be constant given the constant returns to scale production technology.

As shown in Table 3.1, the wage level negotiated between unions and employers often covers far more workers than are actually union members. For this reason, it is assumed here that the wage level resulting from bargaining in the intermediate sector is also paid to (low-skilled) workers in the research and traditional sectors, i.e. $w_{LX} = w_{LR} = w_{LT} = w_L$.[12] With this assumption, the low-skilled reservation wage becomes

$$\bar{w}_{LX} = (1 - u_L)w_L + u_L z \qquad (6.24)$$

Assuming symmetric firms-union pairs so that $w_{L_j} = w_L$ and inserting the above equation into (6.23), leads to equilibrium (nominal) low-skilled wages in the intermediate sector given by

$$w_L^* = \left[1 - \frac{\beta_L(1 - \epsilon_{F,L_L})}{\beta_L(1 - \epsilon_{F,L_L}) + [\beta_L(1 - \epsilon_{F,L_L})\epsilon_{L_j,w} - (\epsilon_{F,L_L} - \omega)(1 - \beta_L)]u_L} \right] z \qquad (6.25)$$

It can be seen from equation (6.25) that an interior solution (i.e. a positive low-skilled wage) only exists if there is low-skilled unemployment, i.e. $u_L > 0$. How high this minimum unemployment rate is, depends on the parameters of the production function and on the degree of union bargaining power. If unemployment falls below this critical level, the (nominal) wage would need

---

[12] Therefore, it is implicitly assumed that this wage is too high to enable all low-skilled workers to take up jobs in either of the other two sectors. It could just as easily be assumed that the wage level in the traditional and research sector is a certain fraction of the intermediate-sector wage. In this case, inter-sectoral wage differentials as in Chapter 4 would occur. However, this assumption would not alter any of the qualitative results derived below.

to fall below the income level guaranteed by unemployment benefits, so that no low-skilled worker would take up a job.

With the labour market now fully specified, it is now possible to derive the general equilibrium. Normalising the price of the final good to unity in all periods and using the fact that in the steady state consumption must grow at the same rate as the rate of endogenous quality improvements, means that the Keynes–Ramsey rule (6.1)

$$r = \rho + \gamma g \tag{6.26}$$

continues to hold, where the growth rate g is defined by equation (6.21) as

$$g = \iota \ln \lambda \tag{6.27}$$

From the household optimisation decisions it is also possible to derive the steady-state demand for the traditional good as

$$T = \frac{1 - \zeta}{p_T} C \tag{6.28}$$

The equilibrium value of demand for the high-tech good M must be equal to the value of output in this sector. This equals total production costs $p_M M$ which are given by the aggregate component costs $p_j X$. This result leads to a (quality adjusted) demand for the manufacturing good given by

$$M = \frac{\zeta}{p_M} C \tag{6.29}$$

For the capital market to be in equilibrium, there cannot be any arbitrage possibilities. With risk-neutral agents this means that the expected return on any shares invested in a research firm must be equal to the rate of return on a riskless asset. In the time interval dt, the shares of every firm currently producing the highest quality intermediate good pay a dividend of $\pi\, dt$. With probability $(1 - \iota\, dt)$, no other firm will successfully innovate during this time interval and replace the current monopolist, in which case the current value of future profits (and thus the price of the shares) increases by $(\dot{V}/\bar{V})\, dt$. However, with probability $\iota\, dt$, another firm will innovate, in which case the new firm charges the limit price and captures all the demand in this market so that the previous monopolist will earn zero profits in the future and thus the capital owners incur a financial loss of $\bar{V}$. Hence, the capital-market equilibrium is characterised by

$$\pi\, dt + \dot{\bar{V}}\, dt(1 - \iota\, dt) - \bar{V}\iota\, dt = rv\, dt$$

Ignoring terms to the power of two and noting that in the steady-state equilibrium $\dot{V} = 0$ leads to

$$\frac{\pi}{V} = r + \iota$$

which from equations (6.17), (6.20), (6.21), (6.26) and (6.29) gives

$$\frac{(1 - 1/\lambda)\zeta}{c_R} = \rho + \iota(1 + \gamma \ln \lambda) \tag{6.30}$$

With unions bargaining for low-skilled wages, the market for low-skilled workers will not clear. Hence, using the labour-demand conditions (6.6), (6.14), and (6.18), means that the low-skilled labour market can be summarised by

$$a_{LR}(w_L, w_H)\iota + a_{LX}(w_L, w_H)X + a_{LT}(w_L, w_H)T = (1 - u_L)(1 - \phi)\bar{L} \tag{6.31}$$

and by analogy, high-skilled labour market equilibrium is characterised by

$$a_{HR}(w_L, w_H)\iota + a_{HX}(w_L, w_H)X + a_{HT}(w_L, w_H)T = \phi\bar{L} \tag{6.32}$$

With the assumptions about the different ratios of low- to high-skilled workers in the different sectors, $a_{HR}/a_{LR} > a_{HX}/a_{LX} > a_{HT}/a_{LT}$ holds. In addition, as $L_{LR} = a_{LR}\iota$ with $\partial a_{LR}/\partial w_L < 0$ and similar relationships hold for the other sectors, it can be easily seen from the low-skilled labour-market condition (6.31) that $du_L/dw_L > 0$. However, whether demand for low-skilled labour increases or decreases when the high-skilled wage increases depends on the elasticity of substitution between the two types of labour, i.e. whether the two types of labour are gross compliments or gross substitutes. Formally, for the case of low-skilled labour, this implies

$$\frac{\partial L_{LS}}{\partial w_H}\bigg|_{w_L, L_H} \gtreqless 0 \text{ for } \sigma_s \gtreqless 1$$

i.e. for $\sigma_s < 1$, the two types of labour are gross compliments and for $\sigma_s > 1$ they are gross substitutes where σ_s denotes the elasticity of substitution between low- and high-skilled labour. From this it follows that

$$\frac{du_L}{dw_H} \gtreqless 0 \text{ for } \sigma_S \lesseqgtr 1 \tag{6.33}$$

Using the definition whereby where $\Theta_{sS} \equiv a_{sS}w_s/c_S$ represents the share of labour of skill-group $s \in \{L, H\}$ in costs incurred in sector $S \in \{R, T, X\}$ as well as the equilibrium demand equations (6.28) together with (6.29) and the price-setting equations (6.5) and (6.16), makes it possible to express the labour-market equations (6.31) and (6.32) in the (quality adjusted) steady state as

$$\Theta_{LR}(w_L, w_H)c_R(w_L, w_H)\iota + \frac{\zeta\Theta_{LX}(w_L, w_H)}{\lambda} + (1-\zeta)\Theta_{LT}(w_L, w_H)$$
$$= w_L(1-u_L)(1-\phi)\bar{L} \quad (6.34)$$

and

$$\Theta_{HR}(w_L, w_H)c_R(w_L, w_H)\iota + \frac{\zeta\Theta_{HX}(w_L, w_H)}{\lambda} + (1-\zeta)\Theta_{HT}(w_L, w_H)$$
$$= w_H\phi\bar{L} \quad (6.35)$$

Equations (6.34) and (6.35) together with the zero-profit condition (6.30) endogenously determine the steady-state innovation rate and factor prices, and by equation (6.25), the equilibrium unemployment rate. This equilibrium is characterised by constant wages in all sectors for both skill groups and positive unemployment and growth rates.

Using the fact that proportional changes in the cost shares can be expressed as $\check{\Theta}_{LS} = \Theta_{HS}(1-\sigma_S)(\check{w}_L - \check{w}_H)$ and $\check{\Theta}_{HS} = -\Theta_{LS}(1-\sigma_S)(\check{w}_L - \check{w}_H)$, where a "check" over a variable denotes the proportional rate of change, e.g. $\check{w}_L \equiv dw_L/w_L$, makes it possible to derive the effects on the research intensity and factor prices associated with changes in the (low-skilled) union bargaining power as[13]

$$\begin{pmatrix} \Phi_{11} & \Phi_{12} & \Phi_{13} \\ \Phi_{21} & \Phi_{22} & \Phi_{23} \\ \Phi_{31} & \Phi_{32} & \Phi_{33} \end{pmatrix} \begin{pmatrix} \check{w}_L \\ \check{w}_H \\ \check{\iota} \end{pmatrix} = \begin{pmatrix} \check{\beta}_L \\ 0 \\ 0 \end{pmatrix} \quad (6.36)$$

where the Φ-coefficients are defined as

$$\Phi_{11} = \frac{1}{\frac{\partial u_L}{\partial \beta_L}(1-\phi)\beta_L\bar{L}}\left[\Theta_{HR}\sigma_R L_{LR} + \Theta_{HX}\sigma_X L_{LX} + \Theta_{HT}\sigma_T L_{LT} + \Theta_{LX}L_{LX}\right.$$
$$\left. + \Theta_{LT}L_{LT} - w_L\frac{\partial u_L}{\partial w_L}(1-\phi)\bar{L}\right]$$

$$\Phi_{12} = -\frac{\Theta_{HR}\sigma_R L_{LR} + \Theta_{HX}\sigma_X L_{LX} + \Theta_{HT}\sigma_T L_{LT} - \Theta_{HX}L_{LX} - \Theta_{HT}L_{LT}}{\frac{\partial u_L}{\partial \beta_L}(1-\phi)\beta_L\bar{L}}$$

$$\Phi_{13} = -\frac{L_{LR}}{\frac{\partial u_L}{\partial \beta_L}(1-\phi)\beta_L\bar{L}}$$

$$\Phi_{21} = \frac{\Theta_{LR}\sigma_R L_{HR} + \Theta_{LX}\sigma_X L_{HX} + \Theta_{LT}\sigma_T L_{HT} - \Theta_{LX}L_{HX} - \Theta_{LT}L_{HT}}{\phi\bar{L}}$$

$$\Phi_{22} = -\frac{\Theta_{LR}\sigma_R L_{HR} + \Theta_{LX}\sigma_X L_{HX} + \Theta_{LT}\sigma_T L_{HT} + \Theta_{HX}L_{HX} + \Theta_{HT}L_{HT}}{\phi\bar{L}}$$

---

[13] See appendices A.8 and A.9 for the derivation.

$$\Phi_{23} = \frac{L_{HR}}{\phi\bar{L}}$$

$$\Phi_{31} = \Theta_{LR}$$

$$\Phi_{32} = \Theta_{HR}$$

$$\Phi_{33} = \frac{\iota(1+\gamma\ln\lambda)}{\rho+\iota(1+\gamma\ln\lambda)}$$

The determinant of the above 3×3 matrix $\mathbf{\Phi}$ is calculated as

$$|\mathbf{\Phi}| = \Phi_{33}(\Phi_{11}\Phi_{22} - \Phi_{12}\Phi_{21}) + \Phi_{32}(\Phi_{13}\Phi_{21} - \Phi_{11}\Phi_{23})$$
$$+ \Phi_{31}(\Phi_{12}\Phi_{23} - \Phi_{13}\Phi_{22})$$

As can be seen from the definition of the Φ-coefficients, this determinant has its absolute maximum for $\sigma_R = \sigma_X = \sigma_T = 0$. Inserting these values and cancelling common terms results in

$$\mathbf{\Phi}_{\min} = \frac{-1}{\frac{\partial u_L}{\partial \beta_L}\beta_L(1-\phi)\phi\bar{L}^2} \left[\frac{\iota(1+\gamma\ln\lambda)}{\rho+\iota(1+\gamma\ln\lambda)} \right.$$
$$((\Theta_{HX}\Theta_{LT} - \Theta_{HT}\Theta_{LX})(L_{HX}L_{LT} - L_{LX}L_{HT})$$
$$- w_L\frac{\partial u_L}{\partial w_L}(1-\phi)\bar{L}(\Theta_{HX}L_{HX} + \Theta_{HT}L_{HT})) + (\Theta_{HR}\Theta_{LX}$$
$$- \Theta_{HX}\Theta_{LR})(L_{HR}L_{LX} - L_{LR}L_{HX}) + (\Theta_{HR}\Theta_{LT} - \Theta_{HT}\Theta_{LR})$$
$$\left. (L_{HR}L_{LT} - L_{LR}L_{HT}) - w_L\frac{\partial u_L}{\partial w_L}(1-\phi)\bar{L}\Theta_{HR}L_{HR} \right] \quad (6.37)$$

As the ratio of high- to low-skilled workers is assumed to be highest in the research and lowest in the traditional sector, $\Theta_{HR} > \Theta_{HX} > \Theta_{HT}, \Theta_{HR} < \Theta_{HX} < \Theta_{HT}$ holds for the relative cost shares and $L_{HR}/L_{HX} > L_{LR}/L_{LX}$ and $L_{HX}/L_{HT} > L_{LX}/L_{LT}$ is true for the relative labour inputs. Therefore and noting from (6.25) that $\partial u_L/\partial w_L < 0$, the terms in the square brackets are all positive and hence the sign of the determinant is unambiguously negative.

Whether higher union bargaining strength has a positive or negative effect on the innovation and hence growth rates can be determined by applying Cramer's rule, which yields,[14]

$$\frac{\breve{\iota}}{\breve{\beta}_L} = \frac{1}{\phi\bar{L}|\mathbf{\Phi}|} \left[\Theta_{HR}(\Theta_{LR}\sigma_R L_{HR} + \Theta_{LX}\sigma_X L_{HX} + \Theta_{LT}\sigma_T L_{HT} - \Theta_{LX}L_{HX} \right.$$
$$- \Theta_{LT}L_{HT}) + \Theta_{LR}(\Theta_{LR}\sigma_R L_{HR} + \Theta_{LX}\sigma_X L_{HX} + \Theta_{LT}\sigma_T L_{HT}$$
$$\left. + \Theta_{HX}L_{HX} + \Theta_{HT}L_{HT}) \right] \quad (6.38)$$

Therefore, whether unions are "good" for growth or not depends on the elasticity of substitution between high- and low-skilled labour. For the extreme

---

[14] See Appendix A.10 for the derivation.

case that these two types of labour are not substitutable, the term in the square brackets reduces to

$$\Theta_{LR}(\Theta_{HX}L_{HX} + \Theta_{HT}L_{HT}) - \Theta_{HR}(\Theta_{LX}L_{HX} + \Theta_{LT}L_{HT})$$

which, with the assumed ranking of the factor intensities, is negative implying a positive effect on the growth rate. The intuition behind this result is that in this case, an increase in bargaining power which increases the low-skilled wage means firms in the manufacturing and traditional sectors are forced to shed some of their (high- and low-skilled) labour. Although the low-skilled wage rate in the research sector also increases, with the ranking of the factor intensities, this sector is able to absorb some of the excess supply of low-skilled labour and will hire additional workers of both skill types. With more researchers employed in the economy, the probability of a successful innovation increases so that the growth rate also rises as a result. However, as soon as a certain degree of substitutability is allowed for in the manufacturing and traditional sectors (a unitary substitution elasticity in these sectors guarantees this result, but depending on the value of σ_R, even lower elasticities of substitution will also be sufficient), the higher low-skilled wage can only be maintained in equilibrium by increasing the productivity of the low-skilled which means attracting some of the high-skilled workers out of the research sector. This in turn has the effect of lowering the total research intensity and hence of reducing the growth rate in the economy.[15]

As shown in Chapter 2, the supply of high-skilled workers has increased substantially in the past two decades. At the same time, particularly the United States and the United Kingdom have witnessed a sharp increase in the wage differential between these two types of labour, whereas in continental Europe relative wages have remained stable but the (low-skilled) unemployment rate has increased dramatically. Using the same techniques as above to derive the effects of an increase in the fraction ϕ of the workforce which is high-skilled leads to

$$\begin{pmatrix} \Delta_{11} & \Delta_{12} & \Delta_{13} \\ \Delta_{21} & \Delta_{22} & \Delta_{23} \\ \Delta_{31} & \Delta_{32} & \Delta_{33} \end{pmatrix} \begin{pmatrix} \breve{w}_L \\ \breve{w}_H \\ \breve{\imath} \end{pmatrix} = \begin{pmatrix} -\breve{\phi} \\ \breve{\phi} \\ 0 \end{pmatrix} \tag{6.39}$$

---

[15] Empirical estimations of the substitution elasticity between low- and high-skilled labour vary greatly. For the U.S., Bound and Johnson (1992) and Katz and Murphy (1992) estimate (absolute) values between 1.4 and 1.7. Estimates for Germany vary a great deal depending on the economic sector and time period which is analysed. For example, whilst Entorf (1996) estimates a value of 1 between low- and high-skilled blue-collar manufacturing workers and between 0.5 and 1.5 for white-collar workers, both Fitzenberger and Franz (1998) and Steiner and Wagner (1998) find values in the range of 0.3 and 0.4. For an overview of these results and more detailed estimates which differentiate between various sectors and between males and females, see Steiner and Mohr (1998).

where the Δ-coefficients are defined as

$$\Delta_{11} = \frac{-1}{(1-u_L)\phi\bar{\bar{L}}}\left[\Theta_{HR}\sigma_R L_{LR}+\Theta_{HX}\sigma_X L_{LX}+\Theta_{HT}\sigma_T L_{LT}+\Theta_{LX} L_{LX}\right.$$

$$\left.+\Theta_{LT}L_{LT}-w_L\frac{\partial u_L}{\partial w_L}(1-\phi)\bar{L}\right]$$

$$\Delta_{12} = \frac{\Theta_{HR}\sigma_R L_{LR}+\Theta_{HX}\sigma_X L_{LX}+\Theta_{HT}\sigma_T L_{LT}-\Theta_{HX}L_{LX}-\Theta_{HT}L_{LT}}{(1-u_L)\phi\bar{L}}$$

$$\Delta_{13} = \frac{L_{LR}}{(1-u_L)\phi\bar{L}}$$

$$\Delta_{21} = \frac{\Theta_{LR}\sigma_R L_{HR}+\Theta_{LX}\sigma_X L_{HX}+\Theta_{LT}\sigma_T L_{HT}-\Theta_{LX}L_{HX}-\Theta_{LT}L_{HT}}{\phi\bar{L}}$$

$$\Delta_{22} = -\frac{\Theta_{LR}\sigma_R L_{HR}+\Theta_{LX}\sigma_X L_{HX}+\Theta_{LT}\sigma_T L_{HT}+\Theta_{HX}L_{HX}+\Theta_{HT}L_{HT}}{\phi\bar{L}}$$

$$\Delta_{23} = \frac{L_{HR}}{\phi\bar{L}}$$

$$\Delta_{31} = \Theta_{LR}$$

$$\Delta_{32} = \Theta_{HR}$$

$$\Delta_{33} = \frac{\iota(1+\gamma\ln\lambda)}{\rho+\iota(1+\gamma\ln\lambda)}$$

Comparing these coefficients with those for an increase in union bargaining power shows that the $\Delta_{11} - \Delta_{13}$ coefficients now have the opposite sign. Therefore, the determinant of the above matrix Δ, now reaches an absolute minimum for $\sigma_R = \sigma_X = \sigma_T = 0$ and is determined as

$$|\Delta| = (\Phi_{11}\Phi_{22}-\Phi_{12}\Phi_{21})\frac{\iota(1+\gamma\ln\lambda)}{\rho+\iota(1+\gamma\ln\lambda)}+\Theta_{HR}(\Phi_{21}\Phi_{13}-\Phi_{11}\Phi_{23})$$

$$+\Theta_{LR}(\Phi_{12}\Phi_{23}-\Phi_{22}\Phi_{13})$$

which is unambiguously positive. Using Cramer's rule, the effect of an increase in the supply of high-skilled workers on low-skilled wages is derived as

$$\frac{\breve{w}_L}{\breve{\phi}} = \frac{1}{\phi\bar{L}|\Delta|}\left[\frac{\iota(1+\gamma\ln\lambda)}{\rho+\iota(1+\gamma\ln\lambda)}\left(\Theta_{LR}\sigma_R L_{HR}+\Theta_{LX}\sigma_X L_{HX}+\Theta_{LT}\sigma_T L_{HT}\right.\right.$$

$$\left.+\Theta_{HX}L_{HX}+\Theta_{HT}L_{HT}\right)+\Theta_{HR}L_{HR}+\frac{1}{1-u_L}\left(\Theta_{HR}L_{LR}\right.$$

$$-\frac{\iota(1+\gamma\ln\lambda)}{\rho+\iota(1+\gamma\ln\lambda)}\left(\Theta_{HR}\sigma_R L_{LR}+\Theta_{HX}\sigma_X L_{LX}+\Theta_{HT}\sigma_T L_{LT}\right.$$

$$\left.\left.\left.-\Theta_{HX}L_{LX}-\Theta_{HT}L_{LT}\right)\right)\right] \quad (6.40)$$

This is unambiguously positive only if $\sigma_R, \sigma_T, \sigma_X \leq 1$. As explained in Chapter 3.2.2, in this case, the two types of labour are gross compliments so that the share of low-skilled labour in unit costs increases as the high-skilled wage falls (which, as shown below, is the case here). Therefore, demand for these workers falls but due to the higher skill intensity and the lower supply of low-skilled workers, their marginal productivity and hence wage is higher than before. For elasticities higher than one, the firms can replace low-skilled workers more easily so that in this case, the fall in low-skilled labour demand outweighs the positive effect of the higher-skill intensity and leads to a fall in their wage level.

The effect of a higher supply of skills on the high-skilled wage is

$$
\frac{\breve{w}_H}{\breve{\phi}} = \frac{1}{\phi \bar{L} |\Delta|} \left[\frac{-1}{1 - u_L} \left(\frac{\iota(1 + \gamma \ln \lambda)}{\rho + \iota(1 + \gamma \ln \lambda)} \left(\Theta_{HR} \sigma_R L_{LR} + \Theta_{HX} \sigma_X L_{LX} \right. \right. \right.
$$

$$
\left. + \Theta_{HT} \sigma_T L_{LT} + \Theta_{LX} L_{LX} + \Theta_{LT} L_{LT} - w_L \frac{\partial u_L}{\partial w_L} (1 - \phi) \bar{L} \right)\right)
$$

$$
+ \Theta_{LR} L_{LR} + \frac{\iota(1 + \gamma \ln \lambda)}{\rho + \iota(1 + \gamma \ln \lambda)} \left(\Theta_{LR} \sigma_R L_{HR} + \Theta_{LX} \sigma_X L_{HX} \right.
$$

$$
\left. \left. + \Theta_{LT} \sigma_T L_{HT} - \Theta_{LX} L_{HX} - \Theta_{LT} L_{HT} \right) - \Theta_{LR} L_{HR} \right] \quad (6.41)
$$

which is unambiguously negative. Hence, as is intuitively to be expected, with the labour market for high-skilled workers clearing, a higher supply of this type of labour leads to fall in their wage rate. As can be seen from the above equation, the presence of union wage bargaining leads to a larger fall in the high-skilled wage rate than would be the case for a competitive market for low-skilled labour. Thus, in accordance with the empirical evidence, the presence of unions means that an increase in the supply of high-skilled labour as witnessed in all industrialised countries in the 1980s and 1990s, compresses the wage differential more in countries with relatively strong unions than in economies where unions have less influence on wages. This is due to the fact that with unions bargaining for the wages of the low-skilled, firms will be more eager to replace these workers with high-skilled counterparts than in a fully competitive labour market. Therefore, the high-skill intensity increases more than in the fully competitive case so that the marginal productivity of high-skilled workers must be lower.

Similarly, the connection between the supply of high-skilled labour and the research intensity (and hence growth rate) is found to be

$$
\frac{\breve{\iota}}{\breve{\phi}} = \frac{1}{\phi \bar{L} |\Delta|} \left[\frac{1}{1 - u_L} \left(\Theta_{HR} \left(\Theta_{HR} \sigma_R L_{LR} + \Theta_{HX} \sigma_X L_{LX} + \Theta_{HT} \sigma_T L_{LT} \right. \right. \right.
$$

$$
\left. + \Theta_{LX} L_{LX} + \Theta_{LT} L_{LT} - w_L \frac{\partial u_L}{\partial w_L} (1 - \phi) \bar{L} \right) + \Theta_{LR} \left(\Theta_{HR} \sigma_R L_{LR} \right.
$$

$$
\left. \left. + \Theta_{HX} \sigma_X L_{LX} + \Theta_{HT} \sigma_T L_{LT} - \Theta_{HX} L_{LX} - \Theta_{HT} L_{LT} \right) \right)
$$

$$- \Theta_{HR}(\Theta_{LR}\sigma_R L_{HR} + \Theta_{LX}\sigma_X L_{HX} + \Theta_{LT}\sigma_T L_{HT} - \Theta_{LX}L_{HX}$$
$$- \Theta_{LT}L_{HT}) - \Theta_{LR}(\Theta_{LR}\sigma_R L_{HR} + \Theta_{LX}\sigma_X L_{HX} + \Theta_{LT}\sigma_T L_{HT}$$
$$+ \Theta_{HX}L_{HX} + \Theta_{HT}L_{HT}) \Big] \quad (6.42)$$

which is always positive. It can be seen from the above equation that the increase in the innovation rate is higher with the presence of unions relative to the fully competitive case. The reason for this is that with low-skilled labour not fully employed, the ratio of high- to low-skilled workers is higher than in the competitive case. With the research sector also being the one that uses high-skilled labour the most intensively, this sector will also profit most from an increase in the fraction of high-skilled workers. It should be noted though, that because some labour remains unemployed, the absolute research intensity and hence the absolute growth rate will be lower in the presence of unions relative to a competitive labour market.

Finally, from equations (6.40) – (6.42) it is possible to analyse the relationship between growth and unemployment caused by an increase in the fraction of workers with a high skill level as

$$\frac{du_L}{d\iota} = \frac{du_L}{dw_L}\frac{dw_L}{d\phi}\frac{d\phi}{d\iota} + \frac{du_L}{dw_H}\frac{dw_H}{d\phi}\frac{d\phi}{d\iota} \quad (6.43)$$

Whether an increase in the supply of high-skilled lowers the unemployment rate for the low-skilled or not depends on the elasticity of substitution between the two types of labour in the three respective sectors. As the higher supply of high-skilled workers leads to a fall in their wage, for "low" values of the substitution elasticities, the low-skilled cost share increases so that firms will hire fewer workers of this type leading to a rise in their unemployment rate. This is the case if there are Cobb–Douglas production technologies in all sectors. For higher values of the substitution elasticities, the low-skilled wage and their cost share fall leading to a reduction in their unemployment rate. However, this effect is counteracted by the fact that the high-skilled wage falls inducing firms to reduce their demand for the low-skilled.

In Chapter 4 it was shown that the degree of product-market power that firms in the manufacturing sector possess, has a positive influence on the (sectoral) unemployment rate. In this chapter, as shown above, the degree of product-market power is determined by the "size" of innovations λ. Using an analogous technique to above, the effect of a higher innovation size is given by

$$\begin{pmatrix} \Gamma_{11} & \Gamma_{12} & \Gamma_{13} \\ \Gamma_{21} & \Gamma_{22} & \Gamma_{23} \\ \Gamma_{31} & \Gamma_{32} & \Gamma_{33} \end{pmatrix} \begin{pmatrix} \breve{w}_L \\ \breve{w}_H \\ \breve{\iota} \end{pmatrix} = \begin{pmatrix} \breve{\lambda} \\ \breve{\lambda} \\ \breve{\lambda} \end{pmatrix} \quad (6.44)$$

where the Γ-coefficients are defined as

$$\Gamma_{11} = -\frac{1}{L_{LX}}\bigg[\Theta_{HR}\sigma_R L_{LR} + \Theta_{HX}\sigma_X L_{LX} + \Theta_{HT}\sigma_T L_{LT} + \Theta_{LX}L_{LX}$$

$$+ \Theta_{LT}L_{LT} - w_L\frac{\partial u_L}{\partial w_L}(1-\phi)\bar{L}\bigg]$$

$$\Gamma_{12} = \frac{\Theta_{HR}\sigma_R L_{LR} + \Theta_{HX}\sigma_X L_{LX} + \Theta_{HT}\sigma_T L_{LT} - \Theta_{HX}L_{LX} - \Theta_{HT}L_{LT}}{L_{LX}}$$

$$\Gamma_{13} = \frac{L_{LR}}{L_{LX}}$$

$$\Gamma_{21} = \frac{\Theta_{LR}\sigma_R L_{HR} + \Theta_{LX}\sigma_X L_{HX} + \Theta_{LT}\sigma_T L_{HT} - \Theta_{LX}L_{HX} - \Theta_{LT}L_{HT}}{L_{HX}}$$

$$\Gamma_{22} = -\frac{\Theta_{LR}\sigma_R L_{HR} + \Theta_{LX}\sigma_X L_{HX} + \Theta_{LT}\sigma_T L_{HT} + \Theta_{HX}L_{HX} + \Theta_{HT}L_{HT}}{L_{HX}}$$

$$\Gamma_{23} = \frac{L_{HR}}{L_{HX}}$$

$$\Gamma_{31} = -\Theta_{LR}\left(\frac{1}{\lambda-1} + \frac{\iota\gamma}{\rho+\iota(1+\gamma\ln\lambda)}\right)^{-1}$$

$$\Gamma_{32} = -\Theta_{HR}\left(\frac{1}{\lambda-1} + \frac{\iota\gamma}{\rho+\iota(1+\gamma\ln\lambda)}\right)^{-1}$$

$$\Gamma_{33} = -\frac{\iota(1+\gamma\ln\lambda)}{\rho+\iota(1+\gamma\ln\lambda)}\left(\frac{1}{\lambda-1} + \frac{\iota\gamma}{\rho+\iota(1+\gamma\ln\lambda)}\right)^{-1}$$

Comparing these coefficients with those for the effect of an increase in bargaining power shows that the determinant of the above 3×3- matrix $|\Gamma|$ has a maximum at $\sigma_R = \sigma_X = \sigma_T = 0$ and is always unambiguously negative. With this knowledge, Cramer's rule can be applied to analyse the effect of a higher innovation size on the low-skilled wage rate as

$$\frac{\breve{w}_L}{\breve{\lambda}} = \frac{1}{|\Gamma|}\Bigg[\left(\frac{1}{\lambda-1} + \frac{\iota\gamma}{\rho+\iota(1+\gamma\ln\lambda)}\right)^{-1}\left(\frac{\iota(1+\gamma\ln\lambda)}{\rho+\iota(1+\gamma\ln\lambda)}\right.$$

$$\left(\frac{1}{L_{HX}}(\Theta_{LR}\sigma_R L_{HR} + \Theta_{LX}\sigma_X L_{HX} + \Theta_{LT}\sigma_T L_{HT} + \Theta_{HX}L_{HX}\right.$$

$$+ \Theta_{HT}L_{HT}) + \frac{1}{L_{LX}}(\Theta_{HR}\sigma_R L_{LR} + \Theta_{HX}\sigma_X L_{LX} + \Theta_{HT}\sigma_T L_{LT}$$

$$\left.\left. - \Theta_{HX}L_{LX} - \Theta_{HT}L_{LT}\right) + \Theta_{HR}\left(\frac{L_{HR}}{L_{HX}} - \frac{L_{LR}}{L_{LX}}\right)\right)$$

$$+ \frac{1}{L_{LX}L_{HX}}\Big(L_{HR}(\Theta_{HR}\sigma_R L_{LR} + \Theta_{HX}\sigma_X L_{LX} + \Theta_{HT}\sigma_T L_{LT}$$

$$- \Theta_{HX}L_{LX} - \Theta_{HT}L_{LT}) + L_{LR}(\Theta_{LR}\sigma_R L_{HR} + \Theta_{LX}\sigma_X L_{HX}$$

$$+ \Theta_{LT}\sigma_T L_{HT} + \Theta_{HX}L_{HX} + \Theta_{HT}L_{HT})\Big)\Bigg] \quad (6.45)$$

Similarly, the effect on the wage rate of the high-skilled is

$$\frac{\breve{w}_H}{\breve{\lambda}} = \frac{1}{|\mathbf{\Gamma}|}\left[\left(\frac{1}{\lambda-1} + \frac{\iota\gamma}{\rho+\iota(1+\gamma\ln\lambda)}\right)^{-1}\left(\frac{\iota(1+\gamma\ln\lambda)}{\rho+\iota(1+\gamma\ln\lambda)}\right.\right.$$

$$\left(\frac{1}{L_{LX}}(\Theta_{HR}\sigma_R L_{LR} + \Theta_{HX}\sigma_X L_{LX} + \Theta_{HT}\sigma_T L_{LT} + \Theta_{LX}L_{LX}\right.$$

$$+\Theta_{LT}L_{LT} - w_L\frac{\partial u_L}{\partial w_L}(1-\phi)\bar{L}) + \frac{1}{L_{HX}}(\Theta_{LR}\sigma_R L_{HR} + \Theta_{LX}\sigma_X L_{HX}$$

$$\left.+ \Theta_{LT}\sigma_T L_{HT} - \Theta_{LX}L_{HX} - \Theta_{LT}L_{HT})\right) - \Theta_{LR}\left(\frac{L_{HR}}{L_{HX}} - \frac{L_{LR}}{L_{LX}}\right)\right)$$

$$+\frac{1}{L_{LX}L_{HX}}\left(L_{LR}(\Theta_{LR}\sigma_R L_{HR} + \Theta_{LX}\sigma_X L_{HX} + \Theta_{LT}\sigma_T L_{HT}\right.$$

$$- \Theta_{LX}L_{HX} - \Theta_{LT}L_{HT}) + L_{HR}(\Theta_{HR}\sigma_R L_{LR} + \Theta_{HX}\sigma_X L_{LX}$$

$$\left.\left.+ \Theta_{HT}\sigma_T L_{LT} + \Theta_{LX}L_{LX} + \Theta_{LT}L_{LT} - w_L\frac{\partial u_L}{\partial w_L}(1-\phi)\bar{L})\right)\right] \quad (6.46)$$

As can be seen from equation (6.45), low-skilled wages increase with higher product-market power if $\sigma_R = \sigma_X = \sigma_T = 0$, but unambiguously decrease as soon as the substitution elasticities reach a value of unity or higher. This can be explained by the fact that the higher innovation size will lead to a fall in the high-skill intensity in the production sectors and to an expansion of the research sector. Therefore, if there are only few substitution possibilities, not only will high-skilled workers leave the production sectors, but as a consequence, low-skilled workers will be dismissed. Therefore, as the production sectors are more low-skill intensive than the research sector, total demand for the low-skilled falls with the remaining workers having a higher marginal productivity than before so that their wage increases. If on the other hand, the two types of labour are "easily" substitutable, then this will lead to a substantial fall in the low-skilled wage so that demand for these workers increases. The wage level of the high-skilled on the other hand, will always decrease. The reason for this is that the increase in λ will shift high-skilled labour from the production sectors into the research sector. Due to this increase in the ratio of high- to low-skilled workers, high-skilled marginal productivity falls. In addition, demand for high-skilled labour in the production sectors decreases. Both effects together lead to a fall in the high-skilled wage level. Thus, how the unemployment rate changes with higher product-market power depends on whether the respective elasticities of substitution are higher than unity or not. If the substitution elasticities are "small", then the fall in low-skilled demand and their higher wage will lead to a rise in unemployment which is slightly counteracted by the rise in low-skilled labour demand in the research sector. The opposite reasoning holds for "high" substitution elasticities. Only for the special case of a Cobb–Douglas production technology in all sectors will the rise in the innovation size and the resulting fall in the low-skilled wage rate unambiguously lead to a fall in the low-skilled unemployment rate.

6.3.1.2 Discussion

This section has integrated union wage negotiations with heterogeneous workers into a quality-ladder model. This makes it possible to analyse the effects of union bargaining power not only on the unemployment rates via their wage-setting effects, but also the effects of unions on the growth rate itself. Firms in the intermediate sector supplying the highest quality good in their market are able to set limit prices and thus earn profits. For this reason, low-skilled workers in this sector are organised in a union in order to redistribute some of the accruing rents. The wage level that results through bilateral bargaining between unions and employers also affects the wages workers in the other sectors demand so that union wage coverage is larger than mere union membership suggests. Within this framework, an increase in the supply of high-skilled labour leads to a higher growth rate because the research sector is the most high-skill labour intensive. The effect on the unemployment rate amongst the low-skilled is unambiguously positive if there is a Cobb–Douglas production technology in all sectors as in this case the share in unit low-skilled-labour costs rises causing firms to demand fewer of these workers but the change in the high-skilled wage leaves the demand for the low-skilled unaltered. For other substitution elasticities the overall effects are ambiguous. Therefore, the result from the last chapter where there was a negative effect of higher growth on the unemployment rate in the context of union wage bargaining, no longer holds in its generality. This can be explained by the fact that growth in the previous chapter was explained exogenously, whereas here it requires low- and high-skilled labour as input.

Whether higher union bargaining power lowers or raises the growth rate again depends on the elasticities of substitution between the two types of labour. If there are no or only very limited substitution possibilities, then the wage increase resulting from an increase in bargaining power will have the effect of directing increased labour of both skill types into the research sector and hence of increasing the growth rate. However, as soon as low-skilled labour becomes more easily substitutable, then the higher low-skilled wage rate will lead firms to raise the skill intensities in the manufacturing and traditional sectors. This can only be achieved by attracting high-skilled labour away from the research sector, thereby lowering the growth rate.

The ambiguity whether unions foster or hinder growth not only hinges on the substitution elasticities, but (in the context of homogeneous labour) on how the engine of growth is modelled. Thus, in a similar model by Stadler (1999), higher union bargaining power always leads to an increase in the growth rate. In his model, this is because the research sector is assumed to be perfectly competitive so that higher union bargaining strength causes a decrease in employment in the intermediate sector and shifts (homogeneous) labour into the research sector, thereby increasing the growth rate. In de Groot (2000, Chap. 6) research is undertaken by intermediate-sector firms themselves. Therefore, a decrease in employment caused by higher wages also reduces the size of the research department and hence has a negative effect on the growth rate. Within an overlapping-generations (OLG) model, Bräuninger (2000) finds that higher union bargaining power causes unemployment to increase and leads to a decrease in the savings of the young so that a negative relationship between unions and growth results. Also within an OLG model, Daveri and Tabellini (2000) show that unemployment caused by unions leads to a decrease in the marginal productivity of capital so that the interest rate also falls. Therefore, workers while they are young will save less which consequently leads to a lower growth rate. However, this result is completely reversed in Irmen and Wigger (2001). In their specification, unions shift income from capital owners to labour which has a higher propensity to save leading to a higher rate of growth.

With unions representing the interests of the low-skilled, the wage rate for this type of labour is above its market-clearing level. Therefore, the wage differential between the two groups is lower the higher the union bargaining power. This effect is further strengthened when the supply of high-skilled labour increases. Further, this higher wage level means that the research intensity and hence the rate of innovations (and growth) is lower than its competitive level. This corresponds to the empirical evidence when comparing the two sides of the atlantic: the United States with weaker unions but a larger wage differential and higher growth rate relative to continental Europe.

Measuring the degree of product-market power by the size of innovations as this determines the markup-factor over marginal costs, then an increase in the innovation size will only reduce the unemployment rate if the substitution elasticities are at least unity, e.g. if there are Cobb–Douglas production technologies in all sectors. In this case, the low-skilled wage rate will decline enough to lead to an increase in the demand for low-skilled labour. If on the other hand there are only limited substitution possibilities between the two types of labour, then the fall in the number of high-skilled workers in the production sectors means that many low-skilled workers need to be dismissed. As the research sector is the least low-skill intensive, this fall in low-skilled labour demand in the production sectors cannot be absorbed by the higher demand for these workers in the research sector so that their unemployment rate must increase. In this latter case, as a higher innovation size with increased research activities always leads to a higher growth rate, growth and (low-skilled) unemployment are positively related.

This chapter so far highlights the fact that the relationship between unions and growth as well as between unemployment and growth is far from unambiguous and depends on the cause of higher growth as well as on the substitution possibilities between low- and high-skilled labour. The next section analyses how unemployment and growth are related in the context of heterogeneous labour and endogenous growth when efficiency-wage considerations are present.

6.3.2 Efficiency Wages

As mentioned in Chapter 3.2, there is substantial evidence that the wages firms pay are motivated by efficiency-wage considerations. Although it is possible to argue that no-shirking, labour-turnover and adverse-selection efficiency wages are all more applicable to highly skilled workers, there are no theoretical reasons why this should be so for efficiency wages based on the notion of a "fair" wage. For this reason, this section assumes that firms will pay fair wages to labour of both skill types.[16]

6.3.2.1 The Model

The setup of the model is very similar to that in the previous sections of this chapter. Thus, the household optimisation decisions are identical to those described in 6.1.1, i.e. infinitely living households gain utility from the consumption of a traditional and a manufacturing good so that the Keynes–Ramsey rule (6.1) as well as the equilibrium expenditure shares (6.3) and (6.4) all continue to hold.

The production of the goods is also similar to that described in Section 6.1.2. However, now, although production is still assumed to take place at constant returns to scale with a constant elasticity of substitution between low- and high-skilled labour as inputs, the production and hence cost functions are also assumed to include the amount of effort workers are exerting as arguments. Hence, analysing the traditional sector first, although perfect competition is still assumed in this sector so that price is equal to marginal cost, the cost function is now given by

$$p_T = c_T(\tilde{e}_{LT}, w_{LT}, \tilde{e}_{HT}, w_{HT}), \qquad \frac{\partial c_T}{\partial \tilde{e}_{sT}} < 0, \frac{\partial c_T}{\partial w_{sT}} > 0 \qquad (6.47)$$

where $\tilde{e}_{LT}$ and $\tilde{e}_{HT}$ are the effort supplies of low- and high-skill workers in the traditional sector respectively, described in more detail below. However, although the effort function also depends on the level of unemployment benefits and the unemployment rate itself, firms treat these variables as exogenous, so that firms only optimise over the wage and employment levels. Hence, the above equation can be written in reduced form as

$$p_T = c_T(w_{LT}, w_{HT}) \qquad (6.48)$$

Shephard's lemma can again be used to determine the respective labour demands as given by equations (6.6) for the low-skilled and (6.7) for the high-skilled although there is now an additional cost reducing effect due to a higher effort supply associated with an increase in the own skill-type wage.

---

[16] See van Schaik and de Groot (1998) and Stadler (1999) for a similar model but with homogeneous workers.

The setup in the manufacturing sector is also similar to that described in
Chapter 6.1.2. Thus, the manufacturing good is produced using a (given)
number of intermediate goods which are constantly being improved due to
successful research activities undertaken in a separate research sector. Therefore, the equations (6.9) – (6.21) all hold (for given effort levels) and solely
the equilibrium wage and effort levels and the resulting general equilibrium
need to be derived.

As throughout in the context of efficiency wages, effort supply is assumed to
be dependent on fair-wage considerations, i.e.

$$\tilde{e}_{sS} = \tilde{e}_{sS}\left(\frac{w_{sS}}{w_{sS}^e}, u_s, z\right), \qquad \frac{\partial \tilde{e}_{sS}}{\partial \frac{w_{sS}}{w_{sS}^e}} > 0, \qquad \frac{\partial \tilde{e}_{sS}}{\partial u_s} > 0, \qquad \frac{\partial \tilde{e}_{sS}}{\partial z} < 0 \qquad (6.49)$$

where $\tilde{e}_{sS}$ is the effort supply, w_{sS} the wage and w_{sS}^e the expected or reference
wage of a worker of type $s \in \{L, H\}$ employed in sector $S \in \{T, X, R\}$.
However, as there is no intuitive reason why workers in different sectors should
have different notions of fairness, the sector index can be omitted. Further,
for the same reason as in Chapter 5.2.2.1, the above specification of the effort
function leads to an unemployment rate which is solely determined by the
parameters of the effort functions and the level of unemployment benefits.[17]
Therefore, whether the wage differential between these two types of labour
is higher or lower than it would be in a competitive labour market depends
on which unemployment rate is higher. For the more plausible case that the
low-skilled have a higher unemployment rate, the wage differential will be less
than in the competitive case.

As both the parameters of the effort function as well as unemployment benefits are exogenously given, as in the previous chapter when wages are set by
efficiency-wage considerations, there is no connection between the growth and
unemployment rates. However, this result crucially rests on the assumption
that the effort supply is only based on within-group wage comparisons. For
this reason, this assumption will now be dropped and instead it is assumed
that workers base their effort decisions depending both on the wage for their

---

[17] Specifying the effort function as in footnote 6 on page 105 but assuming that
the parameters $\varepsilon_1, \varepsilon_2$ and ε_3 are all skill dependent leads to unemployment rates
given by

$$u_s = \frac{z(\varepsilon_{s2} - \varepsilon_{s1})}{\varepsilon_{s3}\varepsilon_{s2}}$$

which means that the assumptions $\varepsilon_{s3} \geq z(\varepsilon_{s2} - \varepsilon_{s1})/\varepsilon_{s2}$ as well as $\varepsilon_{L2} > \varepsilon_{L1} >$
$\varepsilon_{H2} > \varepsilon_{H1}$ and $\varepsilon_{L3} < \varepsilon_{H3}$ are required to ensure that the unemployment rate
for the low-skilled is higher than that of the high-skilled. This corresponds to the
case where low-skilled workers place a higher value on fairness considerations, i.e.
an increase in the relative low-skilled wage leads to a relatively higher increase in
their effort supply.

own as well as the other skill group.[18] Therefore, assuming that workers of the same skill type have the same notion of "fairness" irrespective of the sector they are employed in, means that the above effort functions (6.49) become

$$\tilde{e}_s = \tilde{e}_s\left(w_L, w_H, u_s, z\right), \qquad \frac{\partial \tilde{e}_s}{\partial u_s} > 0, \qquad \frac{\partial \tilde{e}_s}{\partial z} < 0 \qquad (6.50)$$

and

$$\frac{\partial \tilde{e}_L}{\partial w_L} > 0, \frac{\partial \tilde{e}_L}{\partial w_H} < 0, \frac{\partial \tilde{e}_H}{\partial w_L} < 0, \frac{\partial \tilde{e}_H}{\partial w_H} > 0$$

i.e. a higher relative high-skilled wage differential motivates the high-skilled and has the opposite effect on the low-skilled. Further, it is assumed that effort supply is more responsive to changes in the wage rate of one's own skill group. If this were not the case, then a firm would run the risk that the increase in the wage rate of one skill group leads to such a large drop in the effort supply of the other skill group that it is no longer profitable to hire both types of labour.

Profit-maximising firms will choose to set both the wage and employment levels so that the Solow condition is satisfied for both low- and high-skilled workers. Further, as both types of labour are freely mobile across the sectors, the equilibrium relative wage level must be identical across all firms. Therefore, from the Solow condition it is possible to derive an equilibrium relationship between the low- and high-skilled unemployment rates on the one hand and the two wage levels on the other given by

$$u_L = u_L(w_L, w_H, \hat{z}), \qquad \frac{\partial u_L}{\partial w_L} < 0, \frac{\partial u_L}{\partial w_H} > 0, \frac{\partial u_L}{\partial z} > 0, \qquad (6.51)$$

and

$$u_H = u_H(w_L, w_H, z), \qquad \frac{\partial u_H}{\partial w_L} > 0, \frac{\partial u_H}{\partial w_H} < 0, \frac{\partial u_H}{\partial \hat{z}} > 0 \qquad (6.52)$$

This can be explained by the fact that an increase in, for example, the low-skilled unemployment rate leads c.p. to a higher effort supply of the low-skilled so that firms can reduce the low-skilled wage in order to achieve the same effort levels as before the increase in unemployment.[19] As above, which unemployment rate is higher depends on the precise specification of the respective effort

---

[18] In Grossman (2000), depending on the sector in which workers are employed, some workers perceive their wage to be fair based on within skill-group whilst others base their decisions on across-group comparisons. However, the motivation for this assumption is not stated.

[19] To ensure that the Solow condition leads to optimal effort, the elasticity of low-skilled (high-skilled) effort with respect to the low-skilled (high-skilled) wage must be declining. If this were not the case, an increase in the wage rate would lead to a more than proportional increase in effort so that firms could produce the same output with less labour, i.e. $\partial u_s/\partial w_s > 0$. However, in this case, it would be optimal for the firm to pay an infinite wage.

functions. However, again assuming that the low-skilled place more weight on "fairness" considerations will lead to a higher unemployment rate for these workers as a higher low-skilled wage is required to achieve a given effort level.

The research sector is as in Chapter 6.2 so that the equilibrium for this sector is still given by equation (6.30). However, as efficiency-wage considerations are prevalent in both labour markets, neither of these will clear. For this reason, the equilibrium labour-market conditions (6.34) and (6.35) from the previous chapter need to be modified slightly to

$$
\Theta_{LR}(w_L, w_H) c_R(w_L, w_H) \iota + \frac{\varsigma \Theta_{LX}(w_L, w_H)}{\lambda} + (1 - \varsigma)\Theta_{LT}(w_L, w_H)
$$
$$
= w_L(1 - u_L)(1 - \phi)\bar{L} \quad (6.53)
$$

and

$$
\Theta_{HR}(w_L, w_H) c_R(w_L, w_H) \iota + \frac{\varsigma \Theta_{HX}(w_L, w_H)}{\lambda} + (1 - \varsigma)\Theta_{HT}(w_L, w_H)
$$
$$
= w_H(1 - u_H)\phi\bar{L} \quad (6.54)
$$

with the unemployment rates for the low- and high-skilled as defined by (6.51) and (6.52) respectively.

The specification of the labour market fully closes the model so that it is now possible to analyse the comparative-static properties of the general equilibrium. In line with the rest of the chapter, we analyse the effects of an increase in the supply of high-skilled workers as witnessed in the 1980s and 1990s in all industrialised countries. Totally differentiating the credit-market equilibrium equation (6.30) as well as the labour-market equilibrium conditions (6.53) and (6.54) when the fraction ϕ of workers who are high-skilled increases, yields

$$
\begin{pmatrix} \hat{\Delta}_{11} & \hat{\Delta}_{12} & \hat{\Delta}_{13} \\ \hat{\Delta}_{21} & \hat{\Delta}_{22} & \hat{\Delta}_{23} \\ \hat{\Delta}_{31} & \hat{\Delta}_{32} & \hat{\Delta}_{33} \end{pmatrix} \begin{pmatrix} \tilde{w}_L \\ \tilde{w}_H \\ \tilde{\iota} \end{pmatrix} = \begin{pmatrix} -\check{\phi} \\ \check{\phi} \\ 0 \end{pmatrix} \quad (6.55)
$$

where the $\hat{\Delta}$-coefficients are defined as

$$
\hat{\Delta}_{11} = \frac{-1}{(1 - u_L)\phi\bar{L}}\Big[\Theta_{HR}\sigma_R L_{LR} + \Theta_{HX}\sigma_X L_{LX} + \Theta_{HT}\sigma_T L_{LT} + \Theta_{LX}L_{LX}
$$
$$
+ \Theta_{LT}L_{LT} - w_L\frac{\partial u_L}{\partial w_L}(1 - \phi)\bar{L}\Big]
$$

$$
\hat{\Delta}_{12} = \frac{1}{(1 - u_L)\phi\bar{L}}\Big[\Theta_{HR}\sigma_R L_{LR} + \Theta_{HX}\sigma_X L_{LX} + \Theta_{HT}\sigma_T L_{LT} - \Theta_{HX}L_{LX}
$$
$$
- \Theta_{HT}L_{LT} + w_H\frac{\partial u_L}{\partial w_H}(1 - \phi)\bar{L}\Big]
$$

$$
\hat{\Delta}_{13} = \frac{L_{LR}}{(1 - u_L)\phi\bar{L}}
$$

$$\hat{\Delta}_{21} = \frac{1}{(1-u_H)\phi\bar{L}}\left[\Theta_{LR}\sigma_R L_{HR} + \Theta_{LX}\sigma_X L_{HX} + \Theta_{LT}\sigma_T L_{HT} - \Theta_{LX}L_{HX}\right.$$

$$\left. - \Theta_{LT}L_{HT} + w_L\frac{\partial u_H}{\partial w_L}\phi\bar{L}\right]$$

$$\hat{\Delta}_{22} = \frac{-1}{(1-u_H)\phi\bar{L}}\left[\Theta_{LR}\sigma_R L_{HR} + \Theta_{LX}\sigma_X L_{HX} + \Theta_{LT}\sigma_T L_{HT} + \Theta_{HX}L_{HX}\right.$$

$$\left. + \Theta_{HT}L_{HT} - w_H\frac{\partial u_H}{\partial w_H}\phi\bar{L}\right]$$

$$\hat{\Delta}_{23} = \frac{L_{HR}}{(1-u_H)\phi\bar{L}}$$

$$\hat{\Delta}_{31} = \Theta_{LR}$$

$$\hat{\Delta}_{32} = \Theta_{HR}$$

$$\hat{\Delta}_{33} = \frac{\iota(1+\gamma\ln\lambda)}{\rho+\iota(1+\gamma\ln\lambda)}$$

Comparing the above 3×3 matrix $\hat{\Delta}$ with its counterpart (6.39) when wages are set by union wage bargaining shows that its determinant is also always positive.

The change in low-skilled wages caused by a higher fraction of high-skilled workers is derived as

$$\frac{\breve{w}_L}{\phi} = \frac{1}{\phi\bar{L}|\hat{\Delta}|}\left[\frac{1}{1-u_H}\left(\frac{\iota(1+\gamma\ln\lambda)}{\rho+\iota(1+\gamma\ln\lambda)}\left(\Theta_{LR}\sigma_R L_{HR} + \Theta_{LX}\sigma_X L_{HX}\right.\right.\right.$$

$$+ \Theta_{LT}\sigma_T L_{HT} + \Theta_{HX}L_{HX} + \Theta_{HT}L_{HT} - w_H\frac{\partial u_H}{\partial w_H}\phi\bar{L}\Big)$$

$$\left. + \Theta_{HR}L_{HR}\right) + \frac{1}{1-u_L}\left(\Theta_{HR}L_{LR} - \frac{\iota(1+\gamma\ln\lambda)}{\rho+\iota(1+\gamma\ln\lambda)}\left(\Theta_{HR}\sigma_R L_{LR}\right.\right.$$

$$+ \Theta_{HX}\sigma_X L_{LX} + \Theta_{HT}\sigma_T L_{LT} - \Theta_{HX}L_{LX} - \Theta_{HT}L_{LT}$$

$$\left.\left.\left. + w_H\frac{\partial u_L}{\partial w_H}(1-\phi)\bar{L}\right)\right)\right] \quad (6.56)$$

Comparing this effect with that given in equation (6.40) when wages are set by bilateral bargains shows firstly that due to the presence of efficiency-wage considerations amongst the high-skilled, they now also incur unemployment spells and secondly, that there is now an additional term due to the fact that the effort supply of (and hence labour demand for) the low-skilled also (negatively) depends on the high-skilled wage. For this reason, even if the substitution elasticities are below unity, it is not guaranteed that the low-skilled wage will increase as this depends on how the high-skilled wage develops. This change in the high-skilled wage caused by an increase in the number of these workers is calculated as

$$\frac{\breve{w}_H}{\breve{\phi}} = \frac{1}{\phi \bar{L}|\hat{\boldsymbol{\Delta}}|}\left[\frac{-1}{1-u_L}\left(\frac{\iota(1+\gamma\ln\lambda)}{\rho+\iota(1+\gamma\ln\lambda)}\right)\left(\Theta_{HR}\sigma_R L_{LR}+\Theta_{HX}\sigma_X L_{LX}\right.\right.$$

$$+\Theta_{HT}\sigma_T L_{LT}+\Theta_{LX}L_{LX}+\Theta_{LT}L_{LT}-w_L\frac{\partial u_L}{\partial w_L}(1-\phi)\bar{L}\Big)$$

$$+\Theta_{LR}L_{LR}\Big)+\frac{1}{1-u_H}\left(\frac{\iota(1+\gamma\ln\lambda)}{\rho+\iota(1+\gamma\ln\lambda)}\right)\left(\Theta_{LR}\sigma_R L_{HR}\right.$$

$$+\Theta_{LX}\sigma_X L_{HX}+\Theta_{LT}\sigma_T L_{HT}-\Theta_{LX}L_{HX}-\Theta_{LT}L_{HT}$$

$$\left.\left.+w_L\frac{\partial u_H}{\partial w_L}\phi\bar{L}\right)-\Theta_{LR}L_{HR}\right)\right]\quad (6.57)$$

Again, this term is similar to that when unions are present as given by equation (6.41), except that high-skilled unemployment and the effect of the low-skilled wage on the effort supply of the high-skilled need to be taken into account. However, assuming that the change in high-skilled effort supply caused by a change in the low-skilled wage rate is smaller than the change in low-skill effort supply due to a change in their wage rate means that the higher supply of high-skilled workers also leads to a fall in their wage rate as is to be intuitively expected.

Finally, looking at the effects on the innovation rate yields

$$\frac{\breve{\iota}}{\breve{\phi}} = \frac{1}{\phi\bar{\iota}|\hat{\boldsymbol{\Delta}}|}\left[\frac{1}{1-u_L}\left(\Theta_{HR}\left(\Theta_{HR}\sigma_R L_{LR}+\Theta_{HX}\sigma_X L_{LX}+\Theta_{HT}\sigma_T L_{LT}\right.\right.\right.$$

$$+\Theta_{LX}L_{LX}+\Theta_{LT}L_{LT}-w_L\frac{\partial u_L}{\partial w_L}(1-\phi)\bar{L}\Big)+\Theta_{LR}\big(\Theta_{HR}\sigma_R L_{LR}$$

$$+\Theta_{HX}\sigma_X L_{LX}+\Theta_{HT}\sigma_T L_{LT}-\Theta_{HX}L_{LX}-\Theta_{HT}L_{LT}\big)\Big)$$

$$-\frac{1}{1-u_H}\left(\Theta_{HR}\left(\Theta_{LR}\sigma_R L_{HR}+\Theta_{LX}\sigma_X L_{HX}+\Theta_{LT}\sigma_T L_{HT}\right.\right.$$

$$-\Theta_{LX}L_{HX}-\Theta_{LT}L_{HT}+w_L\frac{\partial u_H}{\partial w_L}\phi\bar{L}\Big)+\Theta_{LR}\big(\Theta_{LR}\sigma_R L_{HR}$$

$$+\Theta_{LX}\sigma_X L_{HX}+\Theta_{LT}\sigma_T L_{HT}+\Theta_{HX}L_{HX}+\Theta_{HT}L_{HT}$$

$$\left.\left.\left.-w_H\frac{\partial u_H}{\partial w_H}\phi\bar{L}\right)\right)\right]\quad (6.58)$$

As it is assumed that low-skilled workers effort supply is more responsive to changes in the wage rate than the effort supply of high-skilled workers, the above expression is always positive. Thus, the relationship between the growth and low-skilled unemployment rates due to an increase in the supply of high-skilled workers is similar to that obtained in the presence of bilateral wage bargaining. Due to higher research activities when the supply of high-skilled workers increases, the growth rate will unambiguously increase. For substitution elasticities below unity, low-skilled unemployment will increase as a result of the increased supply of high-skilled. This result is reversed for higher values of the substitution elasticities. However, these results only hold under the assumption that changes in the low-skilled effort supply due to changes in the wage rates are relatively small. Whether the high-skilled unemployment rate rises or falls when the growth rate increases due to a larger fraction of high-skilled workers also depends on the responsiveness of high-skilled effort supply to changes in their wage rate. If these are only relatively small, then the higher supply of these workers leads to a fall in their wage but to a rise in the growth rate as more high-skilled workers are available for research so that growth and unemployment are negatively related for this skill group.

Looking at the effects of higher product-market power, i.e. an increase in the innovation size λ, results in

$$
\begin{pmatrix} \hat{\Gamma}_{11} & \hat{\Gamma}_{12} & \hat{\Gamma}_{13} \\ \hat{\Gamma}_{21} & \hat{\Gamma}_{22} & \hat{\Gamma}_{23} \\ \hat{\Gamma}_{31} & \hat{\Gamma}_{32} & \hat{\Gamma}_{33} \end{pmatrix} \begin{pmatrix} \breve{w}_L \\ \breve{w}_H \\ \breve{\iota} \end{pmatrix} = \begin{pmatrix} \breve{\lambda} \\ \breve{\lambda} \\ \breve{\lambda} \end{pmatrix}
\tag{6.59}
$$

where the $\hat{\Gamma}$-coefficients are defined as

$$
\hat{\Gamma}_{11} = -\frac{1}{L_{LX}}\left[\Theta_{HR}\sigma_R L_{LR} + \Theta_{HX}\sigma_X L_{LX} + \Theta_{HT}\sigma_T L_{LT} + \Theta_{LX} L_{LX} \right.
$$
$$
\left. + \Theta_{LT} L_{LT} - w_L \frac{\partial u_L}{\partial w_L}(1-\phi)\bar{L} \right]
$$

$$
\hat{\Gamma}_{12} = \frac{1}{L_{LX}}\left[\Theta_{HR}\sigma_R L_{LR} + \Theta_{HX}\sigma_X L_{LX} + \Theta_{HT}\sigma_T L_{LT} - \Theta_{HX} L_{LX} \right.
$$
$$
\left. - \Theta_{HT} L_{LT} + w_H \frac{\partial u_L}{\partial w_H}(1-\phi)\bar{L} \right]
$$

$$
\hat{\Gamma}_{13} = \frac{L_{LR}}{L_{LX}}
$$

$$
\hat{\Gamma}_{21} = \frac{1}{L_{HX}}\left[\Theta_{LR}\sigma_R L_{HR} + \Theta_{LX}\sigma_X L_{HX} + \Theta_{LT}\sigma_T L_{HT} - \Theta_{LX} L_{HX} \right.
$$
$$
\left. - \Theta_{LT} L_{HT} + w_L \frac{\partial u_H}{\partial w_L}\phi\bar{L} \right]
$$

$$
\hat{\Gamma}_{22} = -\frac{1}{L_{HX}}\left[\Theta_{LR}\sigma_R L_{HR} + \Theta_{LX}\sigma_X L_{HX} + \Theta_{LT}\sigma_T L_{HT} + \Theta_{HX} L_{HX} \right.
$$
$$
\left. + \Theta_{HT} L_{HT} - w_H \frac{\partial u_H}{\partial w_H}\phi\bar{L} \right]
$$

$$
\hat{\Gamma}_{23} = \frac{L_{HR}}{L_{HX}}
$$

$$
\hat{\Gamma}_{31} = -\Theta_{LR}\left(\frac{1}{\lambda-1} + \frac{\iota\gamma}{\rho+\iota(1+\gamma\ln\lambda)} \right)^{-1}
$$

$$\hat{\Gamma}_{32} = -\Theta_{HR}\left(\frac{1}{\lambda-1} + \frac{\iota\gamma}{\rho+\iota(1+\gamma\ln\lambda)}\right)^{-1}$$

$$\hat{\Gamma}_{33} = -\frac{\iota(1+\gamma\ln\lambda)}{\rho+\iota(1+\gamma\ln\lambda)}\left(\frac{1}{\lambda-1} + \frac{\iota\gamma}{\rho+\iota(1+\gamma\ln\lambda)}\right)^{-1}$$

Comparing the above coefficients with those obtained when wages are set through union wage bargaining shows that the determinant of the above 3 × 3 matrix $|\hat{\Gamma}|$ must have the same sign and is thus always negative. From Cramer's rule, the effect on the low-skilled wage rate is determined as

$$
\frac{\breve{w}_L}{\breve{\lambda}} = \frac{1}{|\hat{\Gamma}|}\left[\left(\frac{1}{\lambda-1} + \frac{\iota\gamma}{\rho+\iota(1+\gamma\ln\lambda)}\right)^{-1}\left(\frac{\iota(1+\gamma\ln\lambda)}{\rho+\iota(1+\gamma\ln\lambda)}\right)\left(\frac{1}{L_{HX}}\right.\right.
$$

$$
\left(\Theta_{LR}\sigma_R L_{HR} + \Theta_{LX}\sigma_X L_{HX} + \Theta_{LT}\sigma_T L_{HT} + \Theta_{HX}L_{HX} + \Theta_{HT}L_{HT}\right.
$$

$$
\left.- w_H \frac{\partial u_H}{\partial w_H}\phi\bar{L}\right) + \frac{1}{L_{LX}}\left(\Theta_{HR}\sigma_R L_{LR} + \Theta_{HX}\sigma_X L_{LX}\right.
$$

$$
\left.+ \Theta_{HT}\sigma_T L_{LT} - \Theta_{HX}L_{LX} - \Theta_{HT}L_{LT} + w_H\frac{\partial u_L}{\partial w_H}(1-\phi)\bar{L}\right)\right)
$$

$$
\left. + \Theta_{HR}\left(\frac{L_{HR}}{L_{HX}} - \frac{L_{LR}}{L_{LX}}\right)\right) + \frac{1}{L_{LX}L_{HX}}\left(L_{HR}\left(\Theta_{HR}\sigma_R L_{LR}\right.\right.
$$

$$
+ \Theta_{HX}\sigma_X L_{LX} + \Theta_{HT}\sigma_T L_{LT} - \Theta_{HX}L_{LX} - \Theta_{HT}L_{LT}
$$

$$
\left.+ w_H\frac{\partial u_L}{\partial w_H}(1-\phi)\bar{L}\right) + L_{LR}\left(\Theta_{LR}\sigma_R L_{HR} + \Theta_{LX}\sigma_X L_{HX}\right.
$$

$$
\left.\left.\left.+ \Theta_{LT}\sigma_T L_{HT} + \Theta_{HX}L_{HX} + \Theta_{HT}L_{HT} - w_H\frac{\partial u_H}{\partial w_H}\phi\bar{L}\right)\right)\right] \quad (6.60)
$$

Similarly, the change in the high-skilled wage is associated with a higher innovation size is given by

$$
\frac{\breve{w}_H}{\breve{\lambda}} = \frac{1}{|\hat{\Gamma}|}\left[\left(\frac{1}{\lambda-1} + \frac{\iota\gamma}{\rho+\iota(1+\gamma\ln\lambda)}\right)^{-1}\left(\frac{\iota(1+\gamma\ln\lambda)}{\rho+\iota(1+\gamma\ln\lambda)}\right.\right.
$$

$$
\left(\frac{1}{L_{LX}}\left(\Theta_{HR}\sigma_R L_{LR} + \Theta_{HX}\sigma_X L_{LX} + \Theta_{HT}\sigma_T L_{LT} + \Theta_{LX}L_{LX}\right.\right.
$$

$$
\left.+ \Theta_{LT}L_{LT} - w_L\frac{\partial u_L}{\partial w_L}(1-\phi)\bar{L}\right) + \frac{1}{L_{HX}}\left(\Theta_{LR}\sigma_R L_{HR} + \Theta_{LX}\sigma_X L_{HX}\right.
$$

$$
\left.+ \Theta_{LT}\sigma_T L_{HT} - \Theta_{LX}L_{HX} - \Theta_{LT}L_{HT} + w_L\frac{\partial u_H}{\partial w_L}\phi\bar{L}\right)\right)
$$

$$
\left.- \Theta_{LR}\left(\frac{L_{HR}}{L_{HX}} - \frac{L_{LR}}{L_{LX}}\right)\right) + \frac{1}{L_{LX}L_{HX}}\left(L_{LR}\left(\Theta_{LR}\sigma_R L_{HR}\right.\right.
$$

$$
\left.+ \Theta_{LX}\sigma_X L_{HX} + \Theta_{LT}\sigma_T L_{HT} - \Theta_{LX}L_{HX} - \Theta_{LT}L_{HT} + w_L\frac{\partial u_H}{\partial w_L}\phi\bar{L}\right)
$$

$$+ L_{HR}\Big(\Theta_{HR}\sigma_R L_{LR} + \Theta_{HX}\sigma_X L_{LX} + \Theta_{HT}\sigma_T L_{LT}$$

$$+ \Theta_{LX}L_{LX} + \Theta_{LT}L_{LT} - w_L \frac{\partial u_L}{\partial w_L}(1 - \phi)\bar{L}\Big)\Big] \quad (6.61)$$

A comparison of the above equations with (6.45) and (6.46) when wages are set by union wage bargaining shows that similar comparative static results occur here, i.e. the wages of the low-skilled increase if there are no or only very limited substitution possibilities and decrease otherwise and the high-skilled wages always fall if the innovation size increases. However, as with efficiency-wage setting firms have more freedom to change the wage rates, the fall in the wage levels is more pronounced (and the increase in the low-skilled wage for "high" substitution elasticities stronger) than is the case with union wage bargaining.

6.3.2.2 Discussion

The above analysis overturns the result from Chapter 5.2.2.1 where there was no relationship between growth and unemployment when wages were set by efficiency-wage considerations. The reason for this is that in contrast to the previous chapter where growth was exogenous, now the parameters of the effort function influence how much labour is devoted to research activities – the source of growth. Whether the relationship between growth and unemployment is positive or negative depends both on whether the focus is on high- or low-skilled workers as well as on the elasticities of substitution between these two types of workers in the different sectors. Thus, if there are only limited substitution possibilities between workers of different skill levels, then even if there is an increase in the supply of high-skilled workers, firms will not be able to easily replace low-skilled workers with high-skilled counterparts. In this case, in each sector the share of low-skilled labour in unit costs increases so that demand for these workers falls and their unemployment rate increases. The opposite result holds for high-skilled workers: both their wage and share in unit costs fall so that their unemployment rate also decreases at the same time as the growth rate increases.

Similar results hold if a higher growth rate is achieved due to higher innovation sizes. In this case there is also the additional effect that resources are shifted towards the research sector which, as this is the most high-skill intensive, leads to a further decrease (increase) in demand for low-skilled (high-skilled) workers.

Thus, in contrast to the previous chapter, the results so far do not vary substantially depending on which type of labour-market imperfection it is that is causing unemployment. The next section analyses the relationship between endogenous growth and unemployment when the labour markets are characterised by matching frictions. However, one drawback of the models so far is that technological progress is not (high-)skill-biased as has been observed. For this reason, the next section slightly alters the above setup to be able to endogenously determine for which skill group innovations are aimed at.

6.3.3 Matching Processes

As shown in Chapter 2, in the 1980s and 90s, wage differentials between high- and low-skilled workers have risen in the United States and United Kingdom and simultaneously there has been a sharp decline in the demand for low-skilled workers. In continental Europe, wage inequality has remained roughly constant but has come at the expense of large increases in the unemployment rate of the low-skilled.[20] Somewhat surprisingly, this rise in relative high-skilled wages and low-skilled unemployment rates was accompanied by a large increase in the supply of skilled labour. Two possible causes for these changes most commonly stated are increased trade with low-skill labour abundant countries (which goes back to Wood 1994) and skilled-biased technological change, (as first formulated by Krugman 1994) whereby new technologies and high-skilled labour are complements. Although most empirical tests tend to favour the skilled-biased technological change hypothesis and tend to dismiss the trade hypothesis (see, for example, Autor et al. 1998, Desjonqueres et al. 1999 and Caselli 1999), as Wood (1998) points out, the two different explanations are not mutually exclusive. Further, Aghion et al. (1999) show that by dropping the assumption that increased international trade is only in final goods and instead analysing increased trade in intermediate goods can account for a much larger share of the increase in wage inequality than the conventional empirical tests. In addition, as shown for example in Caroli and van Reenen (2001), Acemoglu (1999), Kremer and Maskin (1996) and Lindbeck and Snower (1996), there has been a substantial amount of organisational change within firms in recent years and this change has increased the productivity gap (and therefore wages) between workers with different skill levels. However, as shown in Aghion et al. (1999), both the increased trade in intermediate goods as well as the organisational change within firms can be attributed to skilled-biased technological change. It is for this reason that the focus here is on skilled-biased technological change and how it influences wages and the long-run unemployment rates of the high- and low-skilled respectively.

Even growth models developed after the Grossman and Helpman (1991a,b) and Aghion and Howitt (1992, 1994) models presented in this chapter so far

---

[20] For overviews regarding wage differentials, see Katz (2000), Deardorff and Hakura (1994), and Johnson (1997). For empirical evidence on skill-specific unemployment rates see Nickell and Bell (1995, Table 2a, 1996, Table 1), OECD (1997–2002, Table D) and Fitzenberger (1999) for a detailed analysis for West Germany.

either treat labour as homogeneous in which case there cannot be any skill-bias, or, even if heterogeneous labour is taken into account, assume neutral technological change (see, for example, Şener 2001, Li 2001 or Dinopoulos and Thompson 1999). Models that do explicitly incorporate skilled-biased technological change (see, for example, Albrecht and Vroman 2002, Acemoglu 1999, Mortensen and Pissarides 1999c, Gregg and Manning 1997, Mincer 1995, Bound and Johnson 1995, 1992, Katz and Murphy 1992 or Juhn et al. 1993) assume that the cause of the skill bias is exogenous. The effect that an increased supply of skills can generate sufficient technological change to increase the demand for skills was first demonstrated by Eicher (1996) who concentrates on the interaction between human capital accumulation and technological change. In these models, however, the direction of technological progress is exogenous. This is different in the innovation-based growth models by Kiley (1999), who analyses an expanding-variety model, and Acemoglu (1998), who concentrates on rising quality, in which whether new technology is directed at low- or high-skilled workers is endogenously determined. However, all of these models assume perfectly competitive labour markets, so that they cannot analyse the effects of skilled-biased technological change on unemployment. Therefore, in line with the previous models in this chapter, only qualitative growth is investigated and hence Acemoglu (1998) is extended as in Stadler and Wapler (2001) by allowing for matching frictions in the labour market for both low- and high-skilled workers to not only explain the rise in relative high-skilled wages but also in long-term unemployment, especially amongst the low-skilled.

As in Acemoglu (1998), whether research firms develop new components to be used by high- or low-skilled workers depends on two counteracting forces. On the one hand, by the standard substitution effect, a higher supply of skilled labour reduces the high-skilled wage as shown by a downward shift along the short-run labour demand curve L_0^D in Fig. 6.3 from the initial wage differential $(w_H/w_L)_0$ to $(w_H/w_L)_1$. Thus, profits from inventions which are used by the high-skilled decrease as the relative value of high-skilled output falls. For this reason, research firms will want to shift additional resources targeted at inventions to be used by the low-skilled. On the other hand, an increase in the supply of high-skilled labour implies that there are more workers available who are able to use the high-skill complementary components. Therefore, research firms develop new components for a larger market so that the flow profits from these innovations increase. This is called the directed technology effect and leads to an increase in the relative technological level and hence productivity of the high-skilled. In Fig. 6.3 this implies a shift in the labour demand curve to the right as shown by L_1^D. If this shift is large enough, then the new wage differential $(w_H/w_L)_2$ will be higher than its initial level and the long-run labour demand curve L_{LR}^D is upward sloping in this case. Obviously, an analogous argumentation holds for low-skilled workers. In a steady-state equilibrium, expected profits from research efforts targeted at high- and low-skill

Fig. 6.3: Skill-Biased Technological Change and the Wage Differential

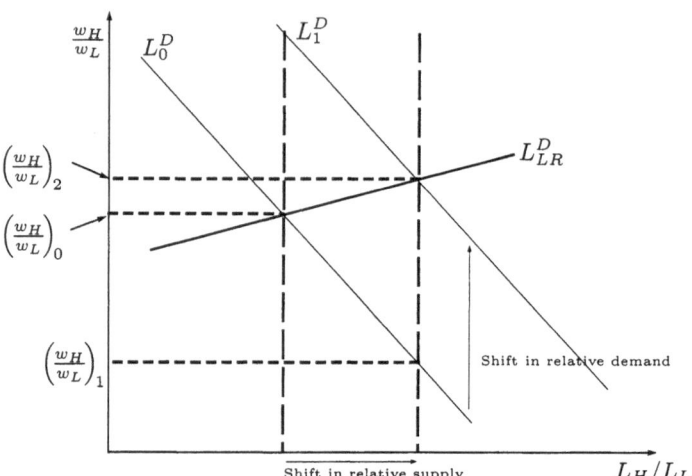

complementary components must be equal. Thus, an increase in the fraction of high-skilled workers can lead both to a lower high-skilled unemployment rate as well as to a larger wage differential between the high- and low-skilled.

6.3.3.1 The Model

The household optimisation decision is as given in Chapter 6.1.1. However, in order to be able to focus on skilled-biased technological change, the production and research sectors need to be specified slightly differently and are similar to that in Acemoglu (1998, 2001a). Aggregate output Y is produced using low- and high-skilled labour according to the CES production function

$$Y = \left[d(A_L L_L^\alpha)^{\frac{\sigma_s - 1}{\sigma_s}} + (1 - d)(A_H L_H^\alpha)^{\frac{\sigma_s - 1}{\sigma_s}} \right]^{\frac{\sigma_s}{\sigma_s - 1}} \tag{6.62}$$

where $d \in (0, 1)$ is a distribution factor measuring the relative importance of the input factors, and $A_s, s \in \{L, H\}$ are the technology parameters of type s labour.

As can be seen by comparing the relative marginal products of high- and low-skilled labour

$$\frac{\partial Y / \partial L_H}{\partial Y / \partial L_L} = \frac{1 - d}{d} \left(\frac{A_H}{A_L} \right)^{\frac{\sigma_s - 1}{\sigma_s}} \left(\frac{L_H}{L_L} \right)^{\frac{\alpha(\sigma_s - 1) - \sigma_s}{\sigma_s}} \tag{6.63}$$

for $\sigma_s > 1$, skilled-biased technological change occurs due to a relative increase in the high-skilled technology A_H which raises the relative marginal productivity of the high-skilled.[21] That is, when the two factors are gross substitutes, an increase in the high-skilled (low-skilled) productivity and the resulting rise (fall) in the relative wage of high- to low-skilled workers leads to a more than proportionate increase in low-skilled (high-skilled) labour demand, so that the relative marginal productivity of the high-skilled will be higher (lower) than before.

In order to keep the analysis tractable, the production function as given by (6.62) is reinterpreted to assume that the consumption good Y is manufactured from two intermediates, each produced in a separate sector using only one type of labour.[22] In this case, $Y = \left[d Y_L^{\frac{\sigma_s - 1}{\sigma_s}} + (1-d) Y_H^{\frac{\sigma_s - 1}{\sigma_s}} \right]^{\frac{\sigma_s}{\sigma_s - 1}}$ where Y_L and Y_H are intermediate goods produced using only low- and high-skilled labour respectively, according to the production functions

$$Y_s = A_s L_s^\alpha \tag{6.64}$$

Denoting with p_s the price of the two intermediate goods means that their relative price can be expressed as

$$\frac{p_H}{p_L} = \frac{1-d}{d} \left(\frac{Y_L}{Y_H} \right)^{\frac{1}{\sigma_s}} \tag{6.65}$$

As each type of labour is only employed in the production of one of these two intermediates, s is both a skill as well as a sector index. Within each sector, production uses sector-specific labour and a continuum $\tilde{j}_s \in [0,1]$ of different components. These components represent the capital stock, but in contrast to the previous chapter, this capital depreciates immediately once a higher quality component has been developed. The assumption that these components are also sector-specific is the means by which the technology differs for high- and low-skilled labour. The highest component quality currently available is denoted by $\tilde{q}_s(\tilde{j})$. The demand for each component $\tilde{j}$ used by firm i in sector s is denoted by $x_s(i,\tilde{j})$ and productivity is given by

$$A_s(i) = \frac{1}{1-\alpha} \int_0^1 \tilde{q}_s(\tilde{j}) x_s(i,\tilde{j})^{1-\alpha} d\tilde{j} \tag{6.66}$$

---

[21] Imposing $\left(\frac{A_H}{A_L} \right)^{\frac{\sigma_s - 1}{\sigma_s}} > \frac{d}{1-d} \left(\frac{L_H}{L_L} \right)^{\frac{\sigma_s - \alpha(\sigma_s - 1)}{\sigma_s}}$ ensures that high-skilled marginal productivity is higher than that of the low-skilled. This result can also be achieved by assuming that the high-skilled are more productive than the low-skilled when using low-skilled technology A_L.

[22] Alternatively, the production function can be interpreted either as there being only one good which is produced using low- and high-skilled workers as imperfect substitutes or as a combination of the above two possibilities with the economy being comprised of various sectors each producing goods which are imperfect substitutes for another and in which all sectors employ both types of labour.

which implies that production of the intermediates Y_L and Y_H takes place at constant returns to scale.

It is shown below that firms in the intermediate sector will always choose to buy the currently highest available quality of each component. Therefore, flow profits of firm i in sector s are determined by

$$\pi_s(i,\tilde{j}) = p_s A_s(i) L_s(i)^\alpha - \int_0^1 \chi_s(\tilde{j}) x_s(i,\tilde{j}) \mathrm{d}\tilde{j} - w_s L_s(i) \tag{6.67}$$

where $L_s(i)$ is the amount of labour of type s employed by firm i, with total labour demand in each sector given by $\int_0^{\tilde{n}_s} L_s(i)\mathrm{d}i = L_s$ where $\tilde{n}_s$ is the mass of firms in either sector, and $\chi_s(\tilde{j})$ is the price of a component with quality $\tilde{q}_s(\tilde{j})$ in the respective sector. Note, however, that due to the constant returns to scale production technology, labour demand in each sector is independent of the number of firms.

Using the profit function (6.67), optimal aggregate demand $X_s(\tilde{j})$ for component $\tilde{j}$ in sector s can be derived as

$$X_s(\tilde{j}) = \left(\frac{p_s \tilde{q}_s(\tilde{j}) L_s^\alpha}{\chi_s(\tilde{j})} \right)^{\frac{1}{\alpha}} \tag{6.68}$$

In all industries in either sector, $\tilde{q}_s(\tilde{j})$ units of the final good are needed to manufacture one unit of the state-of-the-art component $\tilde{j}$. Thus, the production costs increase with the components' quality. An industry leader whose technology is assumed to be perfectly protected by an infinitely lived patent, will maximise his profit function

$$\Pi_s(\tilde{j}) = \left[\chi_s(\tilde{j}) - \tilde{q}_s(\tilde{j}) \right] x_s(\tilde{j}) \tag{6.69}$$

with respect to the price $\chi_s(\tilde{j})$. This yields to a price for each component which is a constant markup over its production costs of

$$\chi_s(\tilde{j}) = \frac{\tilde{q}_s(\tilde{j})}{1-\alpha} \tag{6.70}$$

Each innovation carried out in the research sector improves the quality of a component by the exogenously given factor $\lambda > 1$. The size of each quality improvement is the same for all components in both sectors. Imposing $\lambda > (1-\alpha)^{-(1-\alpha)/\alpha}$, i.e. assuming drastic innovations, ensures that firms producing the intermediate goods Y_L and Y_H will prefer buying the highest quality components even if lower quality ones are sold at marginal costs.

With the component prices as given by equation (6.70), each firm in either sector buys $x_s(i,\tilde{j})$ components so that $X_s(\tilde{j}) = [(1-\alpha)p_s L_s^\alpha]^{\frac{1}{\alpha}}$ and by equation (6.66), equilibrium productivity in sector s is given by

$$A_s = (1 - \alpha)^{\frac{1-2\alpha}{\alpha}} Q_s [p_s L_s^\alpha]^{\frac{1-\alpha}{\alpha}} \tag{6.71}$$

where $Q_s \equiv \int_0^1 \tilde{q}_s(\tilde{j}) d\tilde{j}$ denotes the average quality of components used in the intermediate sectors.

As throughout this chapter, the quality of components can be upgraded by a sequence of innovations each of which builds upon its predecessors and is governed by a Poisson process. In order to ensure that there will always be positive research activities for both types of components, there must be decreasing returns to research effort. Otherwise, if there was an increase in the number of high-skilled whereby the incentives to invent for this skill-group increase, at least in the transition phase, there would be no research efforts targeted at the low-skilled which is clearly at odds with reality. Therefore, the component and sector-specific arrival rate $\iota_s(\tilde{j})$, is specified as

$$\iota_s(\tilde{j}) = Z_s(\tilde{j})\varphi(Z_s(\tilde{j})) \tag{6.72}$$

where $Z_s(\tilde{j})$ is R&D input in terms of the final good with $\varphi'(Z_s(\tilde{j})) < 0$ and $\iota_s(\tilde{j})'(Z_s(\tilde{j})) = \varphi(Z_s(\tilde{j})) + Z_s(\tilde{j})\varphi'(Z_s(\tilde{j})) \geq 0$ to account for the fact that it becomes increasingly difficult to further improve components. With research productivity given by $1/\mu$, at a flow cost of $\mu\tilde{q}_s(\tilde{j})Z_s(\tilde{j})dt$ over the time interval dt, each research firm participating in the innovation race in sector s can attain the stock value $\bar{V}_s(\tilde{j})$ of a successful entrepreneur with the leading technology in industry $\tilde{j}$ with probability $\iota_s(\tilde{j})dt$. Thus, free entry into any innovation race leads to the zero-profit conditions

$$\varphi(Z_s(\tilde{j}))\bar{V}_s(\tilde{j}) = \mu\tilde{q}_s(\tilde{j}) \tag{6.73}$$

which hold for all components in either sector.

As each innovation is specific to one particular component, technological progress is not disembodied as in Chapter 5 where all existing capital did not need to be replaced in order to incorporate the latest technological improvements. With embodied technological change, an intermediate firm must first create a new vacancy and invest in the latest component before being able to implement this new technology. As the innovation size is "drastic", once the latest state-of-the-art technology is adopted, firms will no longer want to operate previous generation components so that the existing employment relationships using outdated components are terminated. Due to matching frictions in the labour market, there is an expected delay denoted by $D_s = 1/f(\theta_s)$ to fill such vacancies. To keep the analysis tractable, it is assumed as in Aghion and Howitt, (1994, 1998, Chap. 4) that by the law of large numbers, the time it takes to fill each vacancy is deterministic. Therefore, during the time span D_s, the current components will still be in use. This means that the total time a component of a particular vintage is in operation is independent of the time needed to fill a vacancy, as the search process delays the expected start and end points by the same amount. Once the vacancies have been filled, the demand for all previous vintages drops to zero and the workers using these now obsolete components are dismissed. Letting τ denote the random time interval between two innovations in an industry in a specific sector, then due to the Poisson process, τ is exponentially distributed over an infinite time horizon with parameter $\iota_s(\tilde{j})$. Therefore, the value $\bar{V}_s(\tilde{j})$ of a research firm owning the leading technology $\tilde{q}_s(\tilde{j})$ in sector s is given by

$$\bar{V}_s(\tilde{j}) = e^{-rD_s} \left\{ \int_0^\infty \iota_s(\tilde{j}) e^{-\iota_s(\tilde{j})\tau} \left[\int_0^\tau \Pi_s(\tilde{j}) e^{-rt} dt \right] d\tau \right\}$$

$$= e^{-rD_s} \frac{\Pi_s(\tilde{j})}{r + \iota_s(\tilde{j})} \tag{6.74}$$

As components are specific to one sector and one type of labour, vacancies are also skill specific. As shown in equation (6.71), the productivity of a firm depends on the average quality Q_s of all components used in production. However, as workers are assumed to operate specific components whose productivity is constant during their lifespan, it is necessary to differentiate between the time a match with job-specific productivity is formed and current time, which we denote by $\tilde{t}$ and t, respectively, with $\tilde{t} \leq t$. Further, with all new jobs using the latest technology, the productivity and hence rents of these jobs are increasing at the rate of technological change. Therefore, in order to be able to hold their current workforce, wages need to be constantly renegotiated to ensure that workers do not quit the firm. This means that the expected present-discounted value of a filled position in sector s, W_s^F, solves the equation

$$rW_s^F(\tilde{t},t) = \tilde{y}_s(\tilde{t}) - w_s(\tilde{t},t) + \iota_s(\tilde{j})\left(W_s^V(t) - W_s^F(\tilde{t},t)\right) + \dot{W}_s^F(\tilde{t},t) \tag{6.75}$$

In equilibrium, the value of a vacancy which is equipped with the latest technology must be zero, i.e.

$$rW_s^V(t) = f(\theta_s) W_s^F(t,t) - c\tilde{y}_s(t) = 0 \tag{6.76}$$

Turning to workers, those currently employed have a present-discounted value of future income $\tilde{V}_s^E$ which solves the asset-pricing equation

$$r\tilde{V}_s^E(\tilde{t},t) = w_s(\tilde{t},t) + \iota_s(\tilde{j})\left(\tilde{V}_s^U(t) - \tilde{V}_s^E(\tilde{t},t)\right) + \dot{\tilde{V}}_s^E(\tilde{t},t) \tag{6.77}$$

and the value for those currently unemployed solves

$$r\tilde{V}_s^U(t) = z(t) + f(\theta_s)\theta_s\left(\tilde{V}_s^E(\tilde{t},t) - \tilde{V}_s^U(t)\right) + \dot{\tilde{V}}_s^U(t) \tag{6.78}$$

Once firms and workers meet, they jointly maximise the Nash product. Using the fact that the equilibrium value of a vacancy is zero, leads to a sharing rule for the surplus created by the match given by

$$\tilde{\beta}_s W_s^F(\tilde{t}, t) = (1 - \tilde{\beta}_s)\left(\tilde{V}_s^E(\tilde{t}, t) - \tilde{V}_s^U(t)\right) \tag{6.79}$$

Further, this sharing rule must also hold for the capital gain terms, i.e.

$$\tilde{\beta}_s \dot{W}_s^F(\tilde{t}, t) = (1 - \tilde{\beta}_s)\left(\dot{\tilde{V}}_s^E(\tilde{t}, t) - \dot{\tilde{V}}_s^U(t)\right) \tag{6.80}$$

From equations (6.75) to (6.80) it is possible to derive the wage equation as

$$w_s(\tilde{t}, t) = \beta_s\left(\tilde{y}_s(\tilde{t}) + c\theta_s \tilde{y}_s(t)\right) + (1 - \tilde{\beta}_s)z(t) \tag{6.81}$$

As can be seen from this equation, the wage depends on the time the match was created and continuously increases as outside options permanently improve because all new jobs are created with the latest technology and are therefore more productive. Therefore, firm profits from a given position continuously decrease so that the job is terminated as soon as these become negative. Inserting the wage equation into (6.75) yields

$$(r + \iota_s(\tilde{j}))W_s^F(\tilde{t}, t) = (1 - \tilde{\beta}_s)\tilde{y}_s(\tilde{t}) - \tilde{\beta}_s c\theta_s \tilde{y}_s(t) - (1 - \tilde{\beta}_s)z(t) + \dot{W}_s^F(\tilde{t}, t) \tag{6.82}$$

Firms know that the expected duration of a match is given by $1/\iota_s(\tilde{j})$, the expected time it takes to improve a current state-of-the-art component. Therefore, from above, at any time $t > \tilde{t}$, the optimal value of a job is

$$W_s^F(\tilde{t}, t) = \int_t^{\tilde{t}+1/\iota_s(\tilde{j})} \left[(1-\tilde{\beta}_s)\tilde{y}_s(\tilde{t}) - \tilde{\beta}_s c\theta_s \tilde{y}_s(t) - (1-\tilde{\beta}_s)z(t)\right] e^{-(r+\iota_s(\tilde{j}))(\bar{t}-t)} \, d\bar{t} \tag{6.83}$$

In the optimum, the value of this position at the end of its expected duration $(t = \tilde{t} + 1/\iota_s(\tilde{j}))$ must be zero. Therefore, the optimum is characterised by

$$(1 - \tilde{\beta}_s)\tilde{y}_s(\tilde{t}) - \tilde{\beta}_s c\theta_s \tilde{y}_s(\tilde{t} + 1/\iota_s(\tilde{j})) - (1 - \tilde{\beta}_s)z(\tilde{t} + 1/\iota_s(\tilde{j})) = 0 \tag{6.84}$$

Inserting this result into the wage equation (6.81) shows that $w_s(\tilde{t}, \tilde{t} + 1/\iota_s(\tilde{j})) = \tilde{y}_s(\tilde{t})$, i.e. at the time the new component is introduced, the wage has risen to the value of the worker output.

With unemployment benefits growing at the same rate as outside opportunities and in a steady state all aggregate variables growing at the rate g, (6.84) can be simplified to

$$(1 - \tilde{\beta}_s) - \tilde{\beta}_s c\theta_s e^{g/\iota_s(\tilde{j})} - \frac{(1 - \tilde{\beta}_s)z(\tilde{t})e^{g/\iota_s(\tilde{j})}}{\tilde{y}_s(\tilde{t})} = 0 \tag{6.85}$$

This equation describing job-destruction shows that whether a higher research intensity $\iota_s(\tilde{j})$ leading to a shorter duration in which each component is in use and a higher growth rate g, are associated with a fall in labour-market tightness or not depends on the elasticity of growth with respect to innovative activity. If this elasticity exceeds unity then a higher research intensity must lead to a fall in labour-market tightness and hence in the wage level in order for the match to be profitable for the firm. More realistic, however, is that this elasticity is well below unity so that a higher research intensity leads to a tighter labour market.[23] Further, it can also be seen from (6.85) that with high-skilled productivity higher than that of the low-skilled, the labour market for the high skilled must be tighter, i.e. $\theta_h > \theta_l$.[24]

Inserting (6.85) into the value of a position as given by (6.83) for $t = \tilde{t}$ yields

$$
W_s^F(t,t) = \tilde{y}_s(t) \int\limits_t^{\tilde{t}+1/\iota_s(j)} \left[(1-\tilde{\beta}_s) - \tilde{\beta}_s c\theta_s e^{gt} - \frac{(1-\tilde{\beta}_s)z(\tilde{t})e^{g/\iota_s(j)}}{\tilde{y}_s(\tilde{t})} \right] \times
$$
$$
e^{-(r+\iota_s(j))(\tilde{t}-t)}\mathrm{d}\tilde{t} \quad (6.86)
$$

The value of this integral decreases with labour-market tightness as a tighter labour market leads to higher wages and also falls at higher growth rates because in this case wages and outside opportunities increase at faster rates. Combining the above equation with the equilibrium vacancy condition (6.76), means that the job-creation condition becomes

$$
\int\limits_t^{\tilde{t}+1/\iota_s(j)} \left[(1-\tilde{\beta}_s) - \tilde{\beta}_s c\theta_s e^{gt} - \frac{(1-\tilde{\beta}_s)z(\tilde{t})e^{g/\iota_s(j)}}{\tilde{y}_s(\tilde{t})} \right] e^{-(r+\iota_s(j))(\tilde{t}-t)}\mathrm{d}\tilde{t} = \frac{c}{f(\theta_s)}
$$
$$
(6.87)
$$

As the arrival rate of new innovations $\iota_s(\tilde{j})$ determines both the job-destruction rate and the expected duration of each job-worker pair, the fraction of jobs that survive in the time interval $1/\iota_s(\tilde{j})$ is e^{-1}. Therefore, using the equilibrium labour-market flow conditions to derive an innovation-based Beveridge curve given by

---

[23] See OECD (2000, Chap. 3) for a summary of estimations of this elasticity in various OECD countries which range from 0.05–0.15%.

[24] Theoretically, it is possible that the market for-low skilled workers is tighter if the high-skilled bargaining power is sufficiently higher than that of the low-skilled. However, this is assumed to not be the case here. Further, this result rests in part on the assumption that unemployment benefits are the same for both types of workers. However, even if unemployment benefits are proportional to the last earned wage, in all countries there are social security payments which guarantee a minimum income. Therefore, interpreting the parameter b more realistically as a combination of (fixed) social security and proportional unemployment benefits would not alter any of our results but would complicate the analysis.

$$u_s = \frac{\iota_s(\tilde{j})}{\iota_s(\tilde{j}) + (1 - e^{-1})\theta_s f(\theta_s)} \tag{6.88}$$

means that, as the research intensity is identical for both skill groups (see below), the high-skilled have a lower unemployment rate and need to spend less time searching for a new job.

In a steady-state equilibrium, the value of firms owning the leading technologies will be constant. Combining this with equations (6.68), (6.69), (6.70) and (6.73) means that the value of a leading research firm as given by (6.74) becomes

$$\alpha(1 - \alpha)^{\frac{1-\alpha}{\alpha}} p_s^{\frac{1}{\alpha}} e^{-rD_s} L_s = \mu \left(\frac{r + Z_s(\tilde{j})\varphi(Z_s(\tilde{j}))}{\varphi(Z_s(\tilde{j}))} \right) \tag{6.89}$$

for all $\tilde{j} \in [0, 1]$ and $s = \{L, H\}$. The l.h.s of (6.89) denotes the flow profits for innovating firms from component sales to the intermediate goods producers. If these profits increase, the research effort $Z_s(\tilde{j})$ aimed at component $\tilde{j}$ will also rise. The profits will be higher if the product price is higher or more workers use the new component so that their demand is higher. Given symmetric firms, it can be seen from (6.89) that $Z_s(\tilde{j}) = Z_s$, i.e. the same amount of research effort is aimed at all components within either sector. Seeing as the r.h.s of (6.89) is increasing in research effort Z_s, it follows that the relative research effort Z_H/Z_L is increasing in $(p_H/p_L) \cdot (e^{-rD_H} L_H / e^{-rD_L} L_L)^{\alpha}$. This means that if p_H is high relative to p_L, it is more profitable to develop high-skill-complementary components because their output commands a higher price. From (6.64), (6.65) and (6.71) it is possible to derive

$$\frac{p_H}{p_L} = \left(\frac{1 - d}{d} \right)^{\frac{\alpha \sigma_s}{1+\alpha(\sigma_s-1)}} \left(\frac{Q_H}{Q_L} \right)^{\frac{-\alpha}{1+\alpha(\sigma_s-1)}} \left(\frac{L_H}{L_L} \right)^{\frac{-\alpha}{1+\alpha(\sigma_s-1)}} \tag{6.90}$$

This equation shows the expected price-effect: for $\sigma_s > 1$, an increase in relative high-skilled labour employment L_H/L_L will c.p. lower the relative price of the intermediate good produced by the high-skilled, so that research efforts targeted at low-skill complementary technologies become more attractive. However, this argument does not take the effect on average technology due to an increase in relative high-skilled labour demand into account.

In a steady state, Q_H/Q_L must be constant so that the respective growth rates of the technological levels must be equal which only occurs if research effort is the same for both sectors, i.e. $Z_H = Z_L$. Inserting this result into (6.89) yields

$$\frac{p_H}{p_L} = \left(\frac{e^{-rD_H} L_H}{e^{-rD_L} L_L} \right)^{-\alpha} \tag{6.91}$$

Combining this equation with (6.90) leads to

$$\frac{Q_H}{Q_L} = \left(\frac{1-d}{d}\right)^{\sigma_s} \left(\frac{e^{-rD_H}}{e^{-rD_L}}\right)^{1+\alpha(\sigma_s-1)} \left(\frac{L_H}{L_L}\right)^{\alpha(\sigma_s-1)} \tag{6.92}$$

This equation highlights the directed technology effect: if the two types of labour are gross substitutes, the relative average technological level is an increasing function of relative labour supply, i.e. if there are more workers using components designed for the high-skilled, it will become increasingly profitable to develop such components.

Assuming perfect competition for the consumption good Y leads to

$$1 = d^{\sigma_s} p_L^{1-\sigma_s} + (1-d)^{\sigma_s} p_H^{1-\sigma_s} \tag{6.93}$$

Combining this equation with the pricing equation (6.91) means that the equilibrium research intensity as given by (6.89) can be rewritten as

$$\alpha(1-\alpha)^{\frac{1-\alpha}{\alpha}} \left[d^{\sigma_s}(e^{-rD_L}L_L)^{\alpha(\sigma_s-1)} + (1-d)^{\sigma_s}(e^{-rD_H}L_H)^{\alpha(\sigma_s-1)} \right]^{\frac{1}{\alpha(\sigma_s-1)}}$$

$$= \mu \left(\frac{r + Z^*\varphi(Z^*)}{\varphi(Z^*)} \right) \tag{6.94}$$

where $Z_H = Z_L = Z^*$ is the equilibrium research effort targeted at any component in either sector. In the case of perfect labour markets with no frictions and no unemployment, the term in square brackets would be larger. Thus, it can immediately be seen from (6.94) that the profits from technology sales are lower than they are in the case of perfectly competitive labour markets. These lower profits reduce the incentives to research. Further, as can be seen from (6.26), the interest rate is an increasing function of the growth rate. Therefore, in order for (6.94) to hold with equality, the interest and growth rates must also be lower than in the case where all markets are fully competitive. There are three reasons for this lower growth rate. Firstly, with unemployment there are fewer employees in each sector which lowers innovation incentives. Secondly, the lower number of workers reduces the amount of final output and thus the available research resources. These are pure scale effects. Thirdly, with matching frictions, all successful innovators have to delay the introduction of their technology by the time interval D_s, further reducing the net value of the innovation.

Given the equilibrium research intensity Z^* and the fact that there is a continuum of components, the economy's growth rate g is derived as

$$g^* = \iota^* \ln \lambda \tag{6.95}$$

with

$$\iota^* = Z^* \varphi(Z^*)$$

Thus, both the innovation arrival and growth rates of the economy are governed by the same exogenous factors.

With these results it is now possible to derive the effects of an increase in the supply of high-skilled workers on the respective unemployment rates. As can be seen from equation (6.92), if the two types of labour are gross substitutes, then an increase in the fraction of high-skilled will lead to more research activities being directed at the high-skilled as there is now a larger demand for these components. Therefore, until the new equilibrium is reached, $Z_H > Z_L$ holds. This means that the productivity gap between the low- and high-skilled increases which induces firms to create more vacancies for the high-skilled and fewer for the low-skilled. Further, as in the transition phase the technological growth rate will be higher for high-skilled than for low-skilled components, but unemployment benefits which determine worker quit behaviour only grow at the level of aggregate growth, the profitability of creating new high-skilled positions increases whereas it decreases for the low-skilled. This has the effect of increasing (decreasing) labour-market tightness for the high- (low-)skilled. However, the long-run effect on labour-market tightness for the two skill groups also depends on whether total research efforts are higher or lower than they were before. As can be seen from (6.94), the higher fraction of high-skilled works as an incentive to invest more into research, but the lower fraction of low-skilled is a disincentive. In addition, the increase in the high-skilled vacancy rate described above delays the production start of new components as firms in the high-skilled sector will need to spend a longer time searching for suitable workers. This also reduces the incentives to innovation but again, the opposite holds for the low-skilled so that the final effect on research is ambiguous. Hence, assuming that the net effect on research incentives is outweighed by the increase (decrease) in the high-skilled (low-skilled) job creation means that the new equilibrium is characterised by a decrease in the high-skilled and an increase in the low-skilled unemployment rate. The same mechanisms also hold if the shift towards the high-skilled occurs due to a decrease in the distribution parameter d, i.e. if the high-skilled intermediates gain in importance for the production of the final good.

As shown above, the growth rate positively depends both on the amount of research activities as well as on the size of innovations λ which is also an indicator of the degree of market power that firms possess as it determines the potential for markup pricing. If this parameter increases, then the growth rate will increase as the same number of innovations lead to higher quality improvements. Therefore, the outside opportunities of currently employed workers increase at a faster rate and hence, as can be seen from the job destruction rate (6.85), the labour markets must become less tight as jobs are now terminated sooner. Therefore, the unemployment rates of both skill groups will increase as a result.

Higher unemployment rates for both types of labour will also result if either
the rate of time preference ρ or the elasticity of marginal utility increase as
both these parameters lead to higher interest rates so that future profits are
discounted at a higher rate and hence job durations will decrease.

Turning to the wage differential between the two skill groups, it can be seen
from equation (6.81) that wages for the two skill groups in terms of the final
good are proportional to their respective values of marginal output, $\tilde{y}_s$. In
relative terms, this value is

$$\frac{Y_H}{Y_L} = \frac{p_H}{p_L}\frac{A_H}{A_L}\left(\frac{L_H}{L_L}\right)^{-(1-\alpha)}$$

which, by inserting equations (6.71), (6.91) and (6.92), becomes

$$\frac{Y_H}{Y_L} = \left(\frac{1-d}{d}\right)^{\sigma_s}\left(\frac{e^{-rD_H}}{e^{-rD_L}}\right)^{\alpha(\sigma_s-1)}\left(\frac{L_H}{L_L}\right)^{\alpha(\sigma_s-1)-1} \tag{6.96}$$

which means from the wage equation (6.81) that the wage differential is proportional to

$$\frac{w_H}{w_L} \propto \frac{\tilde{\beta}_H}{\tilde{\beta}_L}\left(\frac{1-d}{d}\right)^{\sigma_s}\left(\frac{e^{-rD_H}}{e^{-rD_L}}\right)^{\alpha(\sigma_s-1)}\left(\frac{L_H}{L_L}\right)^{\alpha(\sigma_s-1)-1}\left(\frac{1+c\theta_H}{1+c\theta_L}\right)$$
$$\tag{6.97}$$

The relationship given by (6.97) shows that an increase in the relative endow-
ment of high-skilled workers can lead to an increase in the wage differential
between high- and low-skill labour. There are two effects of a higher supply of
high-skilled labour which counteract each other. Firstly, there is the standard
substitution effect which decreases the wage differential. Secondly, there is the
directed technology effect, whereby a larger number of high-skilled workers increases the demand for components complementary to these workers and so
alters the direction of technological change. This increases the incentives to
create vacancies for the high-skilled and makes their market tighter. As can be
seen from the wage equation (6.81), higher labour-market tightness increases
the effective worker bargaining power which further leads to an increase in the
high-skilled wage and, by an analogous argumentation, lowers the low-skilled
wage. However, this direct effect of labour-market tightness is slightly counteracted by the fact that firms now need longer to find these workers which
reduces the incentives to create high-skilled positions. Nevertheless, assuming
that the direct effect of labour-market tightness outweighs, the wage differential will rise.[25] This is in accordance with the empirical evidence as shown for

---

[25] However, as higher high-skilled labour-market tightness also exerts upward pressure on the wage differential, the conditions for a rising wage differential are not
as strong as in the Acemoglu (1998) model.

example in Haskel and Slaughter (2002), whereby rising skill premia is due to technological change which is concentrated in high-skill intensive sectors. The directed technology effect is stronger the larger σ_s is, i.e. the closer substitutes the two intermediates are, or the higher α is, i.e. the smaller are the decreasing returns to labour within each sector. As above, these results also hold if the production of the final good uses more high-skilled intermediates, i.e. the distribution parameter d decreases.

Finally, (6.97) also shows that the wage differential depends on the relative bargaining power of the two types of labour. Since low-skilled workers are more likely to be unionised than their high-skilled counterparts and therefore have a higher bargaining power and that this bargaining power is particularly strong in most European countries, will lead to wage differentials in these countries being more compressed as is empirically the case (see Chapter 2).

The parameter μ is a measure of the marginal research costs associated with inventing a new higher quality component. Therefore, an increase in this parameter will mean that temporarily, the costs of inventing a new component, as given by the r.h.s of equation (6.94), will be higher than the returns. This means that the research intensity Z and thereby the innovation arrival rate ι must fall. This means that the job-destruction rate also falls, so that by equation (6.88), the respective unemployment rates will also decline. This in turn increases the demand for components, thereby increasing the returns from innovation. As these costs equally influence the costs of new components in both sectors, no skill-bias will result, so that the relative productivities of the two labour types and thus the wage differential will remain constant.

There are two direct effects of higher values of either the substitution elasticity σ_s or the elasticity of output with respect to labour α. Firstly, as can be seen from equation (6.96), the relative value of high-skilled marginal output will rise so that the wage differential between the two groups also increases. Secondly, the returns to innovating will increase as innovations now yield a higher flow profit. These direct changes lead to the labour market for the high-skilled becoming relatively tighter. Although this means that firms will now need longer to fill their vacancies so that the return to a new innovation will decrease as it will not be implemented for a longer time period, the overall effect is that research intensity rises. This leads to a higher job-destruction rate so that the overall effect on high-skilled unemployment is ambiguous. On the one hand the higher labour market tightness will decrease high-skilled unemployment, on the other hand the higher innovation arrival rate will increase it. However, the low-skilled unemployment rate will unambiguously rise as due to their lower value of marginal output there will be fewer vacancies as well as a higher job-destruction rate.

The final comparative static effect to be analysed is an increase in the costs of posting a vacancy c. As can be seen from the job-destruction condition (6.85), labour-market tightness must fall if hiring costs increase. This lower value of

θ_s means that firms will need a shorter time period to hire new workers once a new component has been innovated. This increases the returns to innovation so that research intensity Z increases. With both fewer vacancies and a higher innovation arrival rate ι, the unemployment rates for both labour types must increase. Seeing as these higher costs effect both sectors equally, the direction of technological change remains unchanged so that the wage differential will also remain unchanged.

6.3.3.2 Discussion

Since the 1980s, the supply of high-skilled labour has increased sharply in all developed economies. At the same time, technological progress has been (high-)skilled-biased, the wage differential between high- and low-skilled workers has increased or unemployment rates have remained roughly constant for the high-skilled but increased sharply for the low-skilled. The above approach is able to simultaneously explain both of these empirical facts endogenously. An increase in high-skilled labour leads to two counteracting effects. On the one hand, a higher supply leads to a lower wage for this skill-group causing firms to create more positions for these workers. At the same time, the higher number of high-skilled means that there are now more workers able to operate high-skilled components so that demand for these increases. Therefore, researchers will increase their research efforts aimed at components used by high-skilled workers so that technological change will be high-skill biased. Although this bias increases the rate of job destruction, under plausible conditions the job-creation effect will dominate so that the unemployment rate of these workers will fall. At the same time, the increased rate of technological change aimed at high-skilled workers will increase their productivity so that the wage differential also rises.

It was also shown that the innovation rate of an economy characterised by imperfect labour markets is lower than in a perfectly competitive economy. There are three reasons for this result. Firstly, with unemployment there are fewer employees in each sector which lowers innovation incentives. Secondly, the lower number of workers reduces the amount of final output and thus the available research resources. Thirdly, with matching frictions, all successful innovators have to delay the introduction date of their technology, further reducing the net value of innovations. Therefore, when it comes to analysing the determinants of growth, it is not only important to incorporate imperfect labour markets but also to note that the relationship between growth and unemployment may well be different for low- and high-skilled labour.

6.4 Summary and Conclusions

This chapter has analysed the effects of different wage-setting schemes on the relationship between unemployment and growth in the context of heterogeneous labour. In contrast to Chapter 5, the source of long-run growth is now endogenously determined and depends on the amount of labour or final goods devoted to R&D-activities which raise the quality of intermediates or components used in production. For this reason, whether the unemployment rate increases or decreases with higher growth levels depends on whether low- or high-skilled labour is being analysed and also on the source of growth. However, common to all cases and also independent of whether neutral or skill-biased technological change is the focus of attention, is the crucial role that elasticity of substitution between the two types of labour plays. The reason for this is that with heterogeneous labour, this elasticity determines whether the share of unit production costs increases or decreases when the wage rate changes and hence whether demand for this type of labour falls or rises. Thus, if the bargaining power of unions representing low-skilled workers increases, then for the extreme case that there are no substitution possibilities, the growth rate of the economy will increase as more labour of both types is devoted to research. However, this result is overturned for higher elasticity values as in this case, firms in all sectors will choose to increase the ratio of high- to low-skilled workers which implies that total research activities need to decrease as more high-skilled labour is required in the production sectors.

In both the union- and efficiency-wage models, a higher growth rate caused by an increase in the fraction of high-skilled workers in the economy will lead to a higher low-skilled unemployment rate for "low" values of the elasticity of substitution. In this case, the high-skilled wage decreases causing firms to want to substitute low-skilled workers with their high-skilled counterparts. As the research sector is the most high-skill intensive, the rise in high-skilled labour demand will be strongest here and hence attract workers from other sectors. With poor substitution possibilities, as soon as high-skilled workers leave the traditional and manufacturing sectors, many low-skilled workers need to be dismissed. Although the research sector will also demand more low-skilled workers, as this sector is the least low-skill intensive, more workers will be dismissed from the other sectors than can be absorbed by the research sector so that their unemployment rate rises.

If wages are determined in the presence of matching frictions, the substitution elasticity also plays a decisive role. In this model the two types of labour use different technologies. For a linear Cobb–Douglas technology with unit elasticities, changes in the relative technological levels leave the relative marginal productivities unaltered. However, for higher values of this elasticity, a relative improvement of the technology used by the high-skilled will also lead to an increase in the relative high-skilled productivity and is therefore classified as (high-)skill-biased technological change. Therefore, an increase in the number of high-skilled workers makes it more attractive to innovate components used by these workers so that innovations will be biased towards this skill group, leading to an increase in their relative marginal productivity and hence to an increase in job-creation for this group. However, as it is now relatively less attractive to innovate components for low-skilled workers, there is an ambiguous effect on total research activities and hence on the growth and job-destruction rate. This in turn means that the effect on the unemployment rates of the two skill groups is also ambiguous. However, it is more likely that the relationship between growth and unemployment is negative for the high-skilled but positive for the low-skilled workers.

As shown in Chapter 4, the degree of product-market competition indirectly influences product and hence labour demand and therefore the unemployment rate. In this chapter, it is the size of the qualitative improvements associated with each innovation which determines the price markup and hence the amount of product-market power that firms possess. Irrespective of whether wages are set through bilateral bargaining between unions and employers or by efficiency-wage considerations, an increase in this parameter leads to a higher low-skilled unemployment rate if the two types of labour are gross compliments, i.e. the elasticities of substitution in the various sector are smaller than one, as in this case, firms are forced to lower their wage costs by dismissing low-skilled workers. However, as firms have more influence on the wage and employment level when efficiency-wage considerations prevail, the increase in unemployment is lower than is the case with union wage bargaining. Thus, it can also be seen that the result from the previous chapter where, in the presence of efficiency wages, growth and unemployment were independent, no longer holds when the growth rate is determined by the amount of labour devoted to research.

In the case of matching frictions, the degree of product-market power leads to higher unemployment rates for both the low- and high-skilled. This can be explained by the fact that higher innovation sizes leading c.p. to a higher growth rate also increase the interest rate. This means that the returns from creating a new vacancy fall as they are discounted at a higher rate. Therefore, firms will close some of their vacancies so that labour-market tightness falls, causing the unemployment rate to increase for both skill groups.

Finally, all models in this chapter have in common that, as fewer resources are devoted to research than is the case with fully competitive labour markets, the growth rate is also lower. In the union- and efficiency-wage models, this is a pure scale effect. When matching frictions are present, there is an additional factor as innovators face a delay before being able to implement the new technology. This delay means that the profits from such an innovation are lower than they otherwise would be which further reduces the incentives to innovate and hence further lowers the growth rate.

7

Summary and Conclusions

Without doubt, the inability to reduce the high unemployment rates presents one of the most pressing problems in many developed economies. No country, not even the United States, has managed to reduce its unemployment rates to the level it was in the late 1960s and early 1970s, a time when the unemployment rates were three percent or lower, rates at which the labour market is nearly "cleared". Since that time, all economies have experienced negative macroeconomic shocks, for example, the oil crisis, in the case of Germany the unification process is without doubt one of the reasons why the German labour market is struggling, and perhaps even the 11th September 2001 can be seen as a turning point in the economic development of many industrialised nations. Although a rise in unemployment is to be expected after such a shock, the disturbing feature is that it has only been possible to a relatively limited extent to reduce the high unemployment rates from their peaks after the respective shocks.

As shown in Chapter 2 which presented empirical evidence on various aspects of labour-market developments, one of the perhaps surprising facts is that even in times of high unemployment, the real wage level did not fall with very few exceptions. As shown in the subsequent chapters, both employees (represented by unions) and employers (in the form of efficiency-wage considerations) have an intrinsic interest at wages that are above their market-clearing levels. On the side of the unions, this interest can be justified by the fact that it is their objective to maximise the utility of their members which is a positive function of the wage rate as is to be intuitively expected. However, the fact that workers unite to form a union means that this union, as a labour-supply monopolist, has a certain degree of bargaining power when it comes to negotiating over the wage level with employers. Unions can thus use this "power" to extract some of the accruing rents in the labour market. Therefore, reducing union power will reduce wage pressure, but also means that rents now accrue more to employers – a result which may well be combined with extreme social upheaval. In addition, as these bargains take place between autonomous unions and employers, the scope for governmental policy actions is limited.

The second part of Chapter 3 showed why employers benefit from wages that are above their market-clearing levels. This can be explained by the fact that by altering the wage, the employer also as a certain amount of control in providing incentives as to how much effort the employee is willing to provide. The more effort a worker is prepared to provide, the higher will be his productivity. Therefore, although paying a higher wage leads to higher costs for the firm, it also leads to benefits in the form of higher output and hence profit levels. Thus, optimising firms no longer minimise their wage costs, but instead set wage and employment levels to minimise their wage costs per efficiency unit. For example, paying higher wages can lead to reduced shirking by the workforce, in which case the firm has lower monitoring costs, or can lead to less turnover amongst the workers so that the firm saves on training costs for newly hired staff. Another mechanism by which paying higher wages leads to higher productivity levels is the "gift-exchange" or fair-wage hypothesis. This is the approach adopted here whereby workers are willing to provide the required effort in exchange for a wage that they feel as being "fair". As stated at the outset, as the worker needs to be relatively close to his workplace, working and living conditions play a far more important role than they do in other markets. For this reason, the wage he receives plays a larger role than simply being a pure pecuniary "service".

The third major labour-market theory analysed throughout is the matching theory. Here it is recognised that the labour market is characterised by many frictions, due for example to agents who are imperfectly informed about the location and arrival of new job opportunities or suitable workers. Hence, firms and workers need to undergo a time consuming and costly search process before a suitable job-worker match can be formed. It is due to these search costs for both parties that rents occur if a match is formed as both parties can save further costs if they decide to match and quit searching for further applicants or jobs. This means that the wage rate is the result of the bargaining process between the firm and the worker and depends more on the size of the surplus than on the unemployment rate. However, both the unemployment and vacancy rate play an indirect role in the bargaining process, as for example from the worker perspective, if there are "many" other unemployed workers simultaneously looking for a job, then the current applicant will require a longer (and hence more costly) search process before finding another job offer. For this reason, a higher unemployment rate will reduce the worker bargaining power and hence indirectly lead to a lower wage. Similarly, the lower the unemployed benefits an individual receives, the higher is the incentive to accept a job offer as the opportunity costs are lower. However, as these benefits are also associated with many positive aspects, for example a worker will not immediately have to give up his home, lifestyle, can spend more time to find a better match as soon as he loses his job, there are limits as to how low such unemployment benefits can be. Nonetheless, the development in most industrialised countries is towards lower benefits which are only paid a shorter period of time. However, as the market frictions are the main cause of unemployment, it is these that need to be primarily tackled and hence governments need to invest into increasing the efficiency of the matching technology, for example by providing an efficient internet job database which offers a comprehensive job and worker search machine instead of having several parallel databases each only containing a fraction of all jobs and job-seekers. Should individuals or firms not have access to the internet, then local job centres should provide access to this database. In addition, seeing as the incentives to open up a new position are higher the higher worker productivity is, at least in the long-run, unemployment can be reduced through improvements in the education and training system.

As shown in Chapter 4, an imperfectly competitive product market is one of the primary causes of rents accruing to firms. Hence, in the case of unions, the wage a union demands strongly depends on product-market conditions. The reason for this is that more intense competition increases the elasticity of product demand and hence also the wage elasticity of labour demand. Therefore, with more intense competition, unions will reduce their labour demands as otherwise firms would be forced to dismiss so much labour, that unions would not be in their possible utility maximum. These lower wages plus the increased product demand as firms are forced to reduce their prices when competition is more intense, both lead to an increased demand for labour and hence reduce the unemployment rate.

More intensive product-market competition also has positive labour-market effects for the case that efficiency-wage considerations prevail. In this case, even though the wage rate itself does not change as this is still determined by the effort-supply decision, more intense competition and the resulting lower product-market prices lead to an increased demand for these goods, thereby increasing the employment rate. Similarly, in the presence of matching frictions, the higher competition intensity in the product market leads to a fall both in wages and the price that a producer can demand. Again, this leads to consumers increasing their demand for these products and induces firms to create more vacancies which leads to a higher employment. Thus, as it is more plausible that a government has more control over product-market conditions through its anti-trust policies and only has limited scope to intervene in autonomous bargains between firms and employers, the most effective policy recommendation is to reduce as many barriers to product-market competition as possible.

Chapter 5 introduced exogenous growth. As shown in Chapter 2 which provided empirical evidence, the empirical relationship between growth and unemployment is very mixed. This ambiguity also results in the theoretical models. In the case of union wage setting, the higher productivity caused by a rise in the growth rate will lead to a decline in the effective labour costs. It needs to be noted though, that in a steady state with a constant growth rate, unemployment will also remain constant. Similarly, if wages are set by firms according to efficiency-wage calculations, then unemployment and growth are also independent of each other. The reason here is that workers compare their wage to that which similar workers in other firms receive. As technological progress in this chapter was disembodied, the whole capital stock improves with this technological change and hence wages in all firms increase by the same amount, so that the relative wage and hence the effort-supply decision remain unaltered.

In the case of matching frictions, whether the unemployment rate rises or falls with higher growth rates is also unambiguous. On the one hand, a higher growth rate increases the rate at which the costs of posting a vacancy rise in the future. This will make it more attractive for firms to pull forward planned job openings. On the other hand, a higher interest rate also influences the "effective" discount rate as firms profit from the higher capital stock. Whether this rate increases or decreases crucially depends on the intertemporal substitution elasticity as this determines the extent to which the interest rate increases at higher growth levels. For high levels of this elasticity, consumers are relatively willing to have different consumption levels over time and hence, the new equilibrium is characterised by only a small increase in the interest rate so that the effective discount rate decreases. This will increase the value of a position and further induce firms to post additional vacancies. Therefore, in this case, higher growth rates are associated with lower equilibrium unemployment rates. These results, however, are overturned for (empirically more plausible) low values of the intertemporal substitution elasticity in which case the rise in the interest rate causes net returns from posting a vacancy to fall so that fewer jobs will now be created and the unemployment rate increases.

In Chapter 6, many of the features from the previous chapters were combined within one unified setting. Thus, both heterogeneous labour and imperfectly competitive product markets were assumed. In addition, the cause of growth was now no longer assumed to be exogenous, but endogenously determined by innovative activities in a separate research sector which lead to the development of qualitatively higher valued products. As long as the quality-adjusted price of a firm selling the current state-of-the-art intermediates is marginally lower than that of its predecessor, it will be possible for the new incumbent to charge a price over the marginal costs. Therefore, the size of innovative improvements determines the markup factor over the marginal costs and is hence also an indicator of the degree of product power that firms possess. In the case of union wage bargaining or efficiency-wage setting, whether more intense product-market competition leads to a higher or lower unemployment rate depends on the elasticities of substitution between low- and high-skilled labour in the research and production sectors. As a higher innovation size, i.e. less intense product-market competition, increases the incentives to undertake research, additional workers will be attracted to this sector. Further, as the research sector is the most high-skill intensive, this will lead to an increase in demand for this type of labour and attract these workers away from the production sectors. If there are only limited substitution possibilities between the two types of labour, then the fall in the number of high-skilled in the production sectors also leads to a dismissal of low-skilled workers. As the research sector is the low-skilled intensive, these dismissed workers will not all be able to find a job in this sector so that their unemployment rate rises. Thus, as a higher innovation size leads to a higher growth rate, unemployment and growth will be positively correlated for the low-skilled but negatively for the high-skilled as these profit from the expansion of the research sector.

In the case of matching frictions, a higher innovation size leading to an increase in the growth rate will also lead to a corresponding rise in the real interest rate. This in turn has the effect of reducing the incentives to innovate which reduces the rate at which jobs are destroyed. However, job creation also declines because future profits are discounted at a higher rate. This second effect is likely to dominate the lower is the intertemporal elasticity of substitution, as in this case the rise in interest rate will be particularly strong.

As shown in Chapter 2, despite the large increase in the supply of high-skilled workers, the wage differential between low- and high-skilled workers has increased in the United Kingdom and United States but remained more or less constant in France and even decreased in Germany. As shown in Chapter 6.3.1, this development can be explained by the presence of unions who, as empirically plausible, above all represent the interests of low-skilled workers and are particulary strong in Germany and France relative to their influence in the United States and United Kingdom. Similar effects occur in an efficiency-wage setting which builds on the notion of fairness, i.e. where strong disincentives to provide effort arise if the wage differential becomes "too" large.

In the matching model presented in Chapter 6.3.3, research was no longer assumed to be neutral. Instead, the direction of research activities was determined by a substitution and directed technology effect. Thus, an increase in the supply of high-skilled labour leading to a fall in their wage rate and the price of the good these workers produce, will lead more research activities to be targeted at products produced by the low-skilled. However, with a larger number of high-skilled workers, there is a potentially larger demand for products used by this skill group. If this latter effect is strong enough, technological change is high-skill biased as has been observed in recent years and the increase in the number of high-skilled will, in the long-run, also lead to an increase in the wage differential between the two skill groups.

Irrespective of whether wages are set by bilateral bargaining between unions and employers, by efficiency wages or matching frictions, when the growth rate is endogenously determined, the fact that the labour markets do not clear means that labour resources are not used and hence, that the growth rate is lower than it would be in the case of perfectly competitive labour markets. Further, in the case of matching frictions, the lower number of workers reduces the amount of final output and thus the available research resources. Finally, the matching frictions also mean that all successful innovators have to delay the introduction date of their technology so that profits accrue further in the future which additionally reduces the net value of innovations and hence the incentives to innovate. To this extent, government actions aimed at lowering matching frictions will lead to both higher growth rates and lower unemployment rates.

Summing up, it can be seen from the above that even if the government cannot directly interfere in autonomous wage negotiations between unions and employers, policies should be aimed at reducing union bargaining power. This could be done, for example, by making it easier for firms in times of economic downturns or those not belonging to the employer confederation to pay wages which differ from the originally negotiated level. This will have the effect of making the economy more flexible so that the effects of negative shocks can be overcome more quickly. However, it needs to be noted, as can be seen from the efficiency-wage considerations, that low wage levels alone are not a solution. Firms also have an interest at paying higher wages as these play an important role in determining worker motivation and productivity. A policy field where the government does have more scope, is the degree of product-market regulation. A more regulated economy goes at the expense of flexibility if, for example, new firms with better technologies are prevented from entering the market. That a change in policies regulating unions and the product market will not "work" overnight goes without saying, but this also means that expected benefits will also last in the long-term.

Without doubt Germany, but also other industrialised nations, are specialised in producing high quality goods which are high-skilled labour intensive. For this reason, policies must be aimed at increasing the number of workers of this type, again a long-term project. As shown in Chapter 6, such an increase in high-skilled workers changes the incentives to innovate. However, as these incentives are driven by profit-seeking behaviour by firms, policies aimed at reducing this adjustment process, i.e. those which combat the intrinsic interests of firms, will never be successful. Instead, they must aim at making the adjustment process as smooth as possible, for example by providing a comprehensive database containing all job-offerers and job-seekers and by making workers more mobile by forcing them to accept jobs further away from their present location.

Although all these policy considerations have been derived within a closed economy, as stated at the outset, the empirical evidence of the effects on wages and employment of increased trade have found to be small. However, if increased trade has the effect of increasing the intensity of competition, then not only will this also increase employment but again highlights the fact that the economy needs to become more flexible. Seeing as in the past continental European labour markets have been poor at recovering from negative shocks and it is impossible to prevent such shocks from occurring in the future, there is no alternative to increased flexibility.

Thus, although the analysis did not include such features as information asymmetries or uncertainty which without doubt play a role in everyday economic situations but would have made the models even more complex, it can be seen that taken on their own, none of the theories can explain all of the observed unemployment, but together they are able to explain a large proportion. This is all the more plausible as in reality it will never be the case that a labour market is only characterised by one imperfection, but unions, efficiency wages and matching frictions will all be simultaneously present. Now it remains to be seen whether there are politicians willing and able to take the necessary reform steps derived here. Further, hopefully the critical reader has not only been convinced that macroeconomics is good for anything, but also that it is an essential tool to analyse the causes of unemployment and to be able to derive effective policies aimed at reducing the high unemployment rates!

A

Appendix

A.1 Derivation of the Bellman Equation

At any moment in time, the present-discounted value of expected profits is

$$W^V(\tau) = \int_{t=0}^{\tau} -e^{-(f(\theta)+r)t} cy \, dt + e^{-r\tau} \left[e^{-f(\theta)\tau} W^V(\tau) + (1 - e^{-f(\theta)\tau}) W^F(\tau) \right]$$

where the first term on the r.h.s. states that during the time interval τ, the firm has a vacancy which costs cy per unit-time and is filled with probability $f(\theta)$. The second term states that after this time interval has elapsed, then with the probability $e^{-f(\theta)\tau}$ the position is still vacant (and is again expected to be filled at the rate $f(\theta)$ and so has an expected value of W^V), or with the counter-probability it has been filled and has an expected value of W^F. As all these events occur in the future, they need to be discounted at the constant rate r at which the firm could receive capital. Computing the integral leads to

$$W^V(\tau) = \frac{e^{-(f(\theta)+r)\tau} - 1}{f(\theta) + r} cy + e^{-r\tau} \left[e^{-f(\theta)\tau} W^V(\tau) + (1 - e^{-f(\theta)\tau}) W^F(\tau) \right]$$

Solving this expression for $W^V(\tau)$ yields

$$W^V(\tau) = \frac{-cy}{f(\theta) + r} + \frac{e^{-r\tau}(1 - e^{-f(\theta)\tau})}{1 - e^{-(f(\theta)+r)\tau}} W^F(\tau)$$

As the model is set in continuous time, the value of W^V equals the limit of $W^V(\tau)$ as τ approaches to zero. However, seeing as the second term on the r.h.s. of the above equation is not mathematically defined for $\tau = 0$, it is necessary to first apply l'Hôpital's rule and take the first derivative of the nominator and denominator respectively with respect to τ to yield

$$W^V(\tau) = \frac{-cy}{f(\theta) + r} + \frac{-re^{-r\tau}(1 - e^{-f(\theta)\tau}) + e^{-r\tau}f(\theta)e^{-f(\theta)\tau}}{(f(\theta) + r)e^{-(f(\theta)+r)\tau}} W^F(\tau)$$

For $\tau \to 0$, this equation becomes

$$W^V = \frac{-cy}{f(\theta) + r} + \frac{f(\theta)}{f(\theta) + r} W^F \qquad (A.1)$$

Rearranging equation (A.1) yields (3.30) given in the text.

A.2 Derivation of the Relationship between Labour Productivity and Labour-Market Tightness

Equation (3.41) implicitly defines the relationship between labour productivity and labour-market tightness. By the implicit-function theorem, the derivative is given by

$$\frac{d\theta}{dy} = -\frac{f(\theta)[(\rho + \lambda)c - f(\theta)(1 - \tilde{\beta}) + \tilde{\beta}f(\theta)c\theta]}{[\tilde{\beta}[f(\theta)]^2 - (\rho + \lambda)f'(\theta)]cy} \qquad (A.2)$$

As $f'(\theta) < 0$, the denominator is unambiguously positive. The sign of the numerator can be found by using the job-creation condition (3.33) whereby

$$y\left(1 - \frac{\rho + \lambda}{f(\theta)}c\right) = w$$

which, by making use of the wage equation (3.40) implies

$$y\left(1 - \frac{\rho + \lambda}{f(\theta)}c\right) = (1 - \tilde{\beta})z + \tilde{\beta}y(1 + c\theta)$$

which can be rearranged to

$$0 = (\rho + \lambda)c - f(\theta)(1 - \tilde{\beta}) + \tilde{\beta}f(\theta)c\theta + \frac{1 - \tilde{\beta}}{y}zf(\theta)$$

from which it can be seen that $(\rho + \lambda)c - f(\theta)(1 - \tilde{\beta}) + \tilde{\beta}f(\theta)c\theta$ must be negative. Inserting this result into (A.2) shows that labour-market tightness always rises with increases in labour productivity.

A.3 Derivation of the Keynes–Ramsey Rule

Standard dynamic optimisation problems are solved using optimal control theory and the maximum principle. This states that at any time t, the decision or control variable is set so that the state being governed by the control

variable is set optimally. Applying this principle to the dynamic utility maximisation problem yields a five step process. In a first step, the optimisation
problem is formulated which combines the utility function being maximised,
subject to the household dynamic budget constraint, the initial value of the
state variable and the non-negativity constraint to give

$$\max_{C} \int_0^\infty e^{-\rho t} \frac{C_t^{1-\gamma} - 1}{1-\gamma}$$
$$\text{s.t. } \dot{G} = r_t G_t + I_{w_t} - P_t C_t$$
$$G(0) = G_0$$
$$G_t \geq 0$$

From this it is possible to specify the Hamilton function as

$$\mathcal{H} = e^{-\rho t} \frac{C_t^{1-\gamma} - 1}{1-\gamma} + \Lambda(r G_t + I_{w_t} - P_t C_t)$$

In a third step, the Hamilton function is differentiated with respect to the
control variable, here the consumption level, with the optimum being reached
when this derivative is zero. Omitting the time index where no information is
lost by doing so for notational simplicity yields

$$\frac{\partial \mathcal{H}}{\partial C} = e^{-\rho t} C^{-\gamma} - \Lambda P \overset{!}{=} 0$$

from which follows

$$\Lambda = \frac{e^{-\rho t} C^{-\gamma}}{P}$$

and further

$$\frac{\dot{\Lambda}}{\Lambda} = -\rho - \gamma \frac{\dot{C}}{C} - \frac{\dot{P}}{P} \tag{A.3}$$

In a fourth step, the Hamilton function is differentiated with respect to the
state variable, here the stock of assets owned by a household, and this derivative set equal to the negative of the derivative of the Lagrange multiplier
with respect to time which gives

$$\frac{\partial \mathcal{H}}{\partial G} = \Lambda r = -\dot{\Lambda}$$
$$\frac{\dot{\Lambda}}{\Lambda} = -r \tag{A.4}$$

Combining (A.3) and (A.4) results in

$$\frac{\dot{C}}{C} = \frac{1}{\gamma}\left(r - \frac{\dot{P}}{P} - \rho\right) \tag{A.5}$$

Finally, the transversality conditions must hold, which state that

$$\lim_{t \to \infty} \Lambda(t)G(t) = 0$$

which yields the second boundary condition (the first being the initial value of the state variable) for the system to be determinate. Economically, this condition states that a household must take full advantage of all utility-increasing opportunities.

A.4 Derivation of the Price-Index

Equation (4.9) can be rearranged as

$$\frac{M}{T} = \frac{\zeta}{1-\zeta}\frac{p_T}{p_M} \tag{A.6}$$

Inserting this into the consumption function (4.4) yields

$$C = \left(\frac{\zeta}{1-\zeta}\frac{p_T}{p_M}\right)^\zeta T$$

$$CP = Mp_M + Tp_T$$

$$P = \frac{Mp_M + Tp_T}{M^\zeta T^{1-\zeta}}$$

$$= \frac{Mp_M + Tp_T}{\left(\frac{\zeta}{1-\zeta}\frac{p_T}{p_M}\right)^\zeta T}$$

$$= \frac{\frac{M}{T}p_M + p_T}{\left(\frac{\zeta}{1-\zeta}\frac{p_T}{p_M}\right)^\zeta}$$

which by inserting (A.6) from above can be rewritten as

$$P = \frac{\frac{\zeta}{1-\zeta}\frac{p_T}{p_M}p_M + p_T}{\left(\frac{\zeta}{1-\zeta}\frac{p_T}{p_M}\right)^\zeta}$$

$$= \frac{p_T\left(\frac{\zeta}{1-\zeta}+1\right)}{\left(\frac{\zeta}{1-\zeta}\frac{p_T}{p_M}\right)^\zeta}$$

$$= \frac{\left(\frac{p_T}{1-\zeta}\right)}{\left(\frac{p_T}{1-\zeta}\right)^{\zeta}\left(\frac{\zeta}{p_M}\right)^{\zeta}} \tag{A.7}$$

Rearranging equation (A.7) corresponds to the price index as given by (4.11) in the text.

A.5 Derivation of the Demand for a Single Variant when there is Monopolistic Competition

Households want to minimise their consumption expenditures subject to their utility function so that the appropriate Lagrange function is

$$\mathscr{L} = p_{m_i} m_i + \Lambda \left(\left[\sum_1^n m_i^{\kappa} \right]^{\frac{1}{\kappa}} - M \right)$$

Optimising with respect to an arbitrary variant i leads to

$$\frac{\partial \mathscr{L}}{\partial m_i} = p_{m_i} + \Lambda \frac{1}{\kappa} \left[\sum_1^n m_i^{\kappa} \right]^{\frac{1-\kappa}{\kappa}} \sum_1^n \kappa m_i^{\kappa-1} \overset{!}{=} 0 \tag{A.8}$$

and by analogy for a different variant $\tilde{i}$

$$\frac{\partial \mathscr{L}}{\partial m_{\tilde{i}}} = p_{m_{\tilde{i}}} + \Lambda \frac{1}{\kappa} \left[\sum_1^n m_i^{\kappa} \right]^{\frac{1-\kappa}{\kappa}} \sum_1^n \kappa m_{\tilde{i}}^{\kappa-1} \overset{!}{=} 0 \tag{A.9}$$

Equating (A.8) and (A.9) and simplifying yields

$$\frac{p_{m_i}}{m_i^{\kappa-1}} = \frac{p_{m_{\tilde{i}}}}{m_{\tilde{i}}^{\kappa-1}} \tag{A.10}$$

Solving (A.10) for m_i and inserting this into the CES utility function (4.5) yields

$$M = \left[\sum_1^n m_i^{\kappa} \left(\frac{p_{m_i}}{p_{m_{\tilde{i}}}} \right)^{\frac{\kappa}{1-\kappa}} \right]^{\frac{1}{\kappa}}$$

which can be reformulated as

$$M = m_i p_{m_i}^{\frac{1}{1-\kappa}} \left[\sum_1^n p_{m_{\tilde{i}}}^{\frac{\kappa}{\kappa-1}} \right]^{\frac{1}{\kappa}}$$

$$m_i = \frac{p_{m_i}^{\frac{1}{\kappa-1}}}{\left[\sum_1^n p_{m_i}^{\frac{\kappa}{\kappa-1}}\right]^{\frac{1}{\kappa}}} M$$

which by inserting the optimal consumption expenditure share for M as given by (4.9) and making use of the definition of σ, yields

$$m_i = \frac{p_{m_i}^{-\sigma}}{p_M^{-(\sigma-1)}} \zeta P C \tag{A.11}$$

which corresponds to (4.12) as given in the text.

In order to derive the price index for the manufacturing sector, the expenditures per variant are derived by multiplying both sides of (A.11) by p_{m_i} to yield

$$p_i m_i = \frac{p_{m_i}^{1-\sigma}}{\left[\sum_1^n p_{m_i}^{1-\sigma}\right]^{\frac{\sigma}{\sigma-1}}} M$$

The above expression denotes expenditures on the ith variety. Taking sums over all variants gives total manufacturing expenditures as

$$\sum_1^n p_{m_i} m_i = \frac{\sum_1^n p_{m_i}^{1-\sigma}}{\left[\sum_1^n p_{m_i}^{1-\sigma}\right]^{\frac{\sigma}{\sigma-1}}} M$$

Assuming symmetrical firms so that $p_{m_i} = p_{m_i}$ means that the above can be written as

$$\sum_1^n p_{m_i} m_i = \left[\sum_1^n p_i^{1-\sigma}\right]^{\frac{1}{1-\sigma}} M \tag{A.12}$$

where the first term on the r.h.s. of (A.12) is the price index.

A.6 Derivation of the Equilibrium Number of Firms

From the definition of the size of the working population (4.1) and equation (4.31) which gives the number of unemployed, it is possible to derive an equation for the size of the traditional sector as

$$\bar{L}_T = \bar{L} - nL_m \left(1 + \frac{\beta(1-\alpha\kappa)b}{\alpha\kappa(b+\rho)(1-z)}\right) \tag{A.13}$$

Further, using the optimal income shares households spend on either type of good yields

$$\bar{L}_T = \frac{1-\zeta}{\zeta} n p_m m$$

$$= \frac{1-\zeta}{\zeta} n \left(w_M L_m + \tilde{F} \right) \tag{A.14}$$

Equating equations (A.13) and (A.14) and making the appropriate substitutions makes it possible to solve for the number of firms as

$$n = \frac{(b+\rho)(1-z)(1-\alpha\kappa)(\alpha\kappa+\beta(1-\alpha\kappa))\zeta\bar{L}}{\tilde{F}\left[(b+\rho)(1-z)[(1-\zeta)(\alpha\kappa+\beta(1-\alpha\kappa))+(\alpha\kappa)^2\zeta]+(1-\alpha\kappa)b\beta\alpha\kappa\zeta\right]} \tag{A.15}$$

A.7 Derivation of the Growth Rate of the Intermediate Index

Due to the limit-pricing strategy, only the highest quality intermediate goods in each sector are demanded. Therefore, the composite index is now given by (6.9) as

$$\ln M = \int_0^1 \ln \left[\lambda^{\bar{m}_j} x_j \right] \mathrm{d}j$$

which can be rewritten as

$$\ln M = \int_0^1 \ln \lambda^{\bar{m}_j} \, \mathrm{d}j + \int_0^1 \ln x_j \, \mathrm{d}j \tag{A.16}$$

If $\psi_j(\bar{m}, \tau)$ denotes the probability that the product j will be improved $\bar{m}$ times during the time interval τ, then by the law of large numbers, $\psi(\bar{m}, \tau)$ also represents the fraction of industries which experience $\bar{m}$ improvements in this time interval. For the economy as a whole, summing up over all goods which have been improved $\bar{m}$ times for all values of $\bar{m}$ (which will be different for each industry), i.e.

$$\int_0^1 \ln \lambda^{\bar{m}_j} \, \mathrm{d}j = \sum_{\bar{m}=0}^{\infty} \psi(\bar{m}, \tau) \ln \lambda^{\bar{m}}$$

$$= \sum_{\bar{m}=0}^{\infty} \bar{m} \psi(\bar{m}, \tau) \ln \lambda \tag{A.17}$$

where the r.h.s. of (A.17) is the product of the expected number of quality improvements times the size of each quality improvement. Using the characteristics of the Poisson distribution function, the expected number of quality improvements is

$$\mathrm{E}\left[\sum_{\bar{m}=0}^{\infty} \bar{m}\psi(\bar{m},\tau)\right] = \sum_{\bar{m}=0}^{\infty} \frac{\bar{m}(\iota\tau)^{\bar{m}}\mathrm{e}^{-\iota\tau}}{\bar{m}!} = \iota\tau \tag{A.18}$$

Inserting this equation into (A.16) leads to an intermediate index M at time t given by

$$\ln M(\tau) = \iota\tau \ln \lambda + \int_0^1 \ln x_j \, dj$$

which means that the intermediate index grows at the rate

$$\frac{\partial \ln M(\tau)}{\partial \tau} = \frac{\dot{M}}{M} = \iota \ln \lambda \tag{A.19}$$

A.8 Derivation of the Relationship between the Proportional Rate of Change of Cost-Shares and the Elasticity of Substitution

As first shown in Jones (1965, p. 560), given the specified constant returns to scale CES production function with its dual cost function, it holds that:

$$\Theta_{LS}\breve{a}_{LS} + \Theta_{HS}\breve{a}_{HS} = 0 \tag{A.20}$$

$$\Theta_{LS} = \frac{a_{LS}w_L}{c_S} \tag{A.21}$$

$$\Theta_{LS}\breve{w}_L + \Theta_{HS}\breve{w}_H = \breve{c}_S \tag{A.22}$$

further, the elasticity of substitution between high- and low-skilled labour can be written as

$$\sigma_S = \frac{d\left(\frac{H_S}{L_S}\right)\frac{w_L}{w_H}}{d\left(\frac{w_L}{w_H}\right)\frac{H_S}{L_S}}$$

$$= \frac{\frac{da_{HS}}{a_{HS}} \Big/ \frac{da_{LS}}{a_{LS}}}{\frac{dw_L}{w_L} \Big/ \frac{dw_H}{w_H}}$$

$$= \frac{\breve{a}_{HS}/\breve{a}_{LS}}{\breve{w}_L/\breve{w}_H}$$

$$= \frac{\breve{a}_{HS} - \breve{a}_{LS}}{\breve{w}_L - \breve{w}_H} \tag{A.23}$$

From (A.20) we obtain:

$$\breve{a}_{LS} = -\frac{\Theta_{HS}\breve{a}_{HS}}{\Theta_{LS}}$$

Inserting (A.23) yields

$$\breve{a}_{LS} = -\frac{\Theta_{HS}}{\Theta_{LS}}[(\sigma_S(\breve{w}_L - \breve{w}_H) + \breve{a}_{LS}]$$

$$\breve{a}_{LS}\left(1 + \frac{\Theta_{HS}}{\Theta_{LS}}\right) = -\frac{\Theta_{HS}}{\Theta_{LS}}\sigma_S(\breve{w}_L - \breve{w}_H)$$

$$\breve{a}_{LS} = -\frac{\Theta_{HS}\sigma_S(\breve{w}_L - \breve{w}_H)}{\Theta_{LS} + \Theta_{HS}}$$

$$= \Theta_{HS}\sigma_S(\breve{w}_L - \breve{w}_H) \tag{A.24}$$

From (A.21) we obtain:

$$\check{\Theta}_{LS} = \breve{a}_{LS} + \breve{w}_L - \check{c}_S \tag{A.25}$$

Inserting (A.22) leads to:

$$\check{\Theta}_{LS} = \breve{a}_{LS} + \breve{w}_L - \Theta_{LS}\breve{w}_L - \Theta_{HS}\breve{w}_H$$

$$\breve{a}_{LS} = \check{\Theta}_{LS} - \breve{w}_L + \Theta_{LS}\breve{w}_L + \Theta_{HS}\breve{w}_H$$

$$= \check{\Theta}_{LS} - \breve{w}_L + (1 - \Theta_{HS})\breve{w}_L + \Theta_{HS}\breve{w}_H$$

$$= \check{\Theta}_{LS} - \Theta_{HS}(\breve{w}_L - \breve{w}_H) \tag{A.26}$$

Inserting (A.26) into (A.24) results in:

$$\check{\Theta}_{LS} - \Theta_{HS}(\breve{w}_L - \breve{w}_H) = -\Theta_{HS}\sigma_S(\breve{w}_L - \breve{w}_H)$$

$$\check{\Theta}_{LS} = \Theta_{HS}(1 - \sigma_S)(\breve{w}_L - \breve{w}_H) \tag{A.27}$$

Similarly, from (A.20) we obtain:

$$\breve{a}_{HS} = -\frac{\Theta_{LS}\breve{a}_{LS}}{\Theta_{HS}}$$

Inserting (A.23) yields

$$\breve{a}_{HS} = \frac{\Theta_{LS}}{\Theta_{HS}}[(\sigma_S(\breve{w}_L - \breve{w}_H) - \breve{a}_{HS}]$$

$$\check{a}_{HS}\left(1+\frac{\Theta_{LS}}{\Theta_{HS}}\right) = \frac{\Theta_{LS}}{\Theta_{HS}}\sigma_S(\check{w}_L - \check{w}_H)$$

$$\check{a}_{HS} = \Theta_{LS}\sigma_S(\check{w}_L - \check{w}_H) \tag{A.28}$$

Using the definition

$$\Theta_{HS} = \frac{a_{HS}w_H}{c_S}$$

we obtain:

$$\check{\Theta}_{HS} = \check{a}_{HS} + \check{w}_H - \check{c}_S \tag{A.29}$$

Inserting (A.22) leads to:

$$\check{\Theta}_{HS} = \check{a}_{HS} + \check{w}_H - \Theta_{LS}\check{w}_L - \Theta_{HS}\check{w}_H$$

$$\check{a}_{HS} = \check{\Theta}_{HS} - \check{w}_H + \Theta_{LS}\check{w}_L + \Theta_{HS}\check{w}_H$$

$$= \check{\Theta}_{HS} - \check{w}_H + \Theta_{LS}\check{w}_L + (1 - \Theta_{LS})\check{w}_H$$

$$= \check{\Theta}_{HS} + \Theta_{LS}(\check{w}_L - \check{w}_H) \tag{A.30}$$

Inserting (A.30) into (A.28) results in:

$$\check{\Theta}_{HS} = -\Theta_{LS}(1 - \sigma_S)(\check{w}_L - \check{w}_H) \tag{A.31}$$

A.9 Derivation of the Influence of Union Bargaining Power on the General Equilibrium

Analysing the effects of higher union bargaining power β_L yields from equation (6.34)

$$\frac{\partial \Theta_{LR}}{\partial w_L}\,\mathrm{d}w_L c_R \iota + \frac{\partial c_R}{\partial w_L}\,\mathrm{d}w_L \Theta_{LR}\iota + \frac{\partial \Theta_{LR}}{\partial w_H}\,\mathrm{d}w_H c_R \iota + \frac{\partial c_R}{\partial w_H}\,\mathrm{d}w_H \Theta_{LR}\iota$$

$$+ \Theta_{LR} c_R \,\mathrm{d}\iota + \frac{\zeta}{\lambda}\left[\frac{\partial \Theta_{LX}}{\partial w_L}\,\mathrm{d}w_L + \frac{\partial \Theta_{LX}}{\partial w_H}\,\mathrm{d}w_H\right] + (1-\zeta)\left[\frac{\partial \Theta_{LT}}{\partial w_L}\,\mathrm{d}w_L\right.$$

$$\left. + \frac{\partial \Theta_{LT}}{\partial w_H}\,\mathrm{d}w_H\right] = \left[(1-u_L)(1-\phi)\bar{L} - w_L\frac{\partial u_L}{\partial w_L}(1-\phi)\bar{L}\right]\mathrm{d}w_L$$

$$- \left[\frac{\partial u_L}{\partial \beta_L}w_L(1-\phi)\bar{L}\right]\mathrm{d}\beta_L$$

which can be rearranged to

$$\left[\frac{\partial\Theta_{LR}}{\partial w_L}c_{R\iota} + \frac{\partial c_R}{\partial w_L}\Theta_{LR\iota} + \frac{\zeta}{\lambda}\frac{\partial\Theta_{LX}}{\partial w_L} + (1-\zeta)\frac{\partial\Theta_{LT}}{\partial w_L} - (1-u_L)(1-\phi)\bar{L}\right.$$

$$\left. + w_L\frac{\partial u_L}{\partial w_L}(1-\phi)\bar{L}\right]dw_L + \left[\frac{\partial\Theta_{LR}}{\partial w_H}c_{R\iota} + \frac{\partial c_R}{\partial w_H}\Theta_{LR\iota} + \frac{\zeta}{\lambda}\frac{\partial\Theta_{LX}}{\partial w_H}\right.$$

$$\left. + (1-\zeta)\frac{\partial\Theta_{LT}}{\partial w_H}\right]dw_H + \Theta_{LR}c_R\, d\iota = -\left[\frac{\partial u_L}{\partial\beta_L}w_L(1-\phi)\bar{L}\right]d\beta_L \quad (A.32)$$

Noting that

$$\frac{\partial\Theta_{LS}}{\partial w_L} = \frac{\left(\frac{\partial a_{LS}}{\partial w_L}w_L + a_{LS}\right)c_S - a_{LS}w_L\frac{\partial c_S}{\partial w_L}}{c_S^2}$$

In the optimum, the first term in the numerator will be zero. Therefore, the above expression simplifies to

$$\frac{\partial\Theta_{LS}}{\partial w_L} = a_{LS}\left(\frac{1}{c_S} - \frac{\Theta_{LS}}{c_S}\right)$$

$$= \frac{L_S}{S}\frac{\Theta_{HS}}{c_S} \quad (A.33)$$

and by an analogous derivation

$$\frac{\partial\Theta_{LS}}{\partial w_H} = -\frac{H_S}{S}\frac{\Theta_{LS}}{c_S} \quad (A.34)$$

$$\frac{\partial\Theta_{HS}}{\partial w_L} = -\frac{L_S}{S}\frac{\Theta_{HS}}{c_S} \quad (A.35)$$

$$\frac{\partial\Theta_{HS}}{\partial w_H} = \frac{H_S}{S}\frac{\Theta_{LS}}{c_S} \quad (A.36)$$

From this and using the steady-state demand levels means that (A.32) becomes

$$\left[L_{LR}\Theta_{HR} + L_{LR}\Theta_{LR} + L_{LX}\Theta_{HX} + L_{LT}\Theta_{HT} - (1-u_L)(1-\phi)\bar{L}\right.$$

$$\left. + w_L\frac{\partial u_L}{\partial w_L}(1-\phi)\bar{L}\right]dw_L - \left[L_{HR}\Theta_{LR} - L_{HR}\Theta_{LR} + L_{HX}\Theta_{LX}\right.$$

$$\left. + L_{HT}\Theta_{LT}\right]dw_H + \Theta_{LR}c_R\, d\iota = -\left[\frac{\partial u_L}{\partial\beta_L}w_L(1-\phi)\bar{L}\right]d\beta_L$$

from which follows

$$\left[L_{LR} - (1-u_L)(1-\phi)\bar{L} + w_L\frac{\partial u_L}{\partial w_L}(1-\phi)\bar{L}\right]dw_L$$

$$+ \frac{\zeta}{\lambda}\Theta_{LX}\Theta_{HX}(\breve{w}_L - \breve{w}_H) + (1-\zeta)\Theta_{LT}\Theta_{HT}(\breve{w}_L - \breve{w}_H) + \Theta_{LR}c_R\, d\iota$$

$$= -\left[\frac{\partial u_L}{\partial \beta_L} w_L (1-\phi)\bar{L}\right] \mathrm{d}\beta_L \quad (A.37)$$

From (A.26) it can be seen that in a steady-state setting with $\check{L}_S = 0$

$$\Theta_{HS}(\check{w}_L - \check{w}_H) = \check{\Theta}_{LS}$$

holds, so that (A.37) becomes

$$w_L L_{LR}\check{w}_L + w_L L_{LR}\check{\imath} + w_L L_{LX}\check{\Theta}_{LX} + w_L L_{LT}\check{\Theta}_{LT}$$

$$= \left[(1-u_L)(1-\phi)\bar{L} - w_L\frac{\partial u_L}{\partial w_L}(1-\phi)\bar{L}\right]\mathrm{d}w_L - \left[\frac{\partial u_L}{\partial \beta_L}w_L(1-\phi)\bar{L}\right]\mathrm{d}\beta_L$$

$$w_L L_{LR}(\check{w}_L + \check{\imath}) + w_L L_{LX}\check{\Theta}_{LX} + w_L L_{LT}\check{\Theta}_{LT}$$

$$= \left[(1-u_L)(1-\phi)\bar{L} - w_L\frac{\partial u_L}{\partial w_L}(1-\phi)\bar{L}\right]\mathrm{d}w_L - \left[\frac{\partial u_L}{\partial \beta_L}w_L(1-\phi)\bar{L}\right]\mathrm{d}\beta_L$$

which by using (A.25) can be rewritten as

$$w_L L_{LR}(\check{\Theta}_{LR} + \check{c}_R + \check{\imath}) + w_L L_{LX}\check{\Theta}_{LX} + w_L L_{LT}\check{\Theta}_{LT}$$

$$= \left[(1-u_L)(1-\phi)\bar{L} - w_L\frac{\partial u_L}{\partial w_L}(1-\phi)\bar{L}\right]\mathrm{d}w_L - \left[\frac{\partial u_L}{\partial \beta_L}w_L(1-\phi)\bar{L}\right]\mathrm{d}\beta_L$$

$$\frac{[\check{\Theta}_{LR} + \check{c}_R + \check{\imath}]L_{LR}}{(1-\phi)\beta_L\bar{L}} + \frac{\check{\Theta}_{LX}L_{LX}}{(1-\phi)\beta_L\bar{L}} + \frac{\check{\Theta}_{LT}L_{LT}}{(1-\phi)\beta_L\bar{L}}$$

$$= \left[\frac{1-u_L}{\beta_L} - \frac{w_L}{\beta_L}\frac{\partial u_L}{\partial w_L}\right]\check{w}_L - \frac{\partial u_L}{\partial \beta_L}\check{\beta}_L$$

Inserting (A.22) and (A.27) leads to

$$\frac{1}{(1-\phi)\beta_L\bar{L}}\left[[\Theta_{HR}(1-\sigma_R)(\check{w}_L - \check{w}_H) + \Theta_{LR}\check{w}_L + \Theta_{HR}\check{w}_H + \check{\imath}]L_{LR}\right.$$

$$\left. + [\Theta_{HX}(1-\sigma_X)(\check{w}_L - \check{w}_H)]L_{LX} + [\Theta_{HT}(1-\sigma_T)(\check{w}_L - \check{w}_H)]L_{LT}\right]$$

$$= \left[\frac{1-u_L}{\beta_L} - \frac{w_L}{\beta_L}\frac{\partial u_L}{\partial w_L}\right]\check{w}_L - \frac{\partial u_L}{\partial \beta_L}\check{\beta}_L$$

$$\frac{-\check{w}_L}{(1-\phi)\beta_L\bar{L}}\left[\Theta_{HR}\sigma_R L_{LR} - \Theta_{LR}L_{LR} + \Theta_{HX}\sigma_X L_{LX} + \Theta_{HT}\sigma_T L_{LT}\right.$$

$$\left. -\Theta_{HR}L_{LR} - \Theta_{HX}L_{LX} - \Theta_{HT}L_{LT} + (1-u_L)(1-\phi)\bar{L} - w_L\frac{\partial u_L}{\partial w_L}(1-\phi)\bar{L}\right]$$

$$+ \frac{\check{w}_H}{(1-\phi)\beta_L\bar{L}}\left[\Theta_{HR}\sigma_R L_{LR} + \Theta_{HR}L_{LR} + \Theta_{HX}\sigma_X L_{LX} + \Theta_{HT}\sigma_T L_{LT}\right.$$

$$-\Theta_{HR}L_{LR} - \Theta_{HX}L_{LX} - \Theta_{HT}L_{LT}\Big] + \frac{\iota L_{LR}}{(1-\phi)\beta_L\bar{L}} = -\frac{\partial u_L}{\partial\beta_L}\breve{\beta}_L$$

$$\frac{-\breve{w}_L}{(1-\phi)\beta_L\bar{L}}\Big[\Theta_{HR}\sigma_R L_{LR} + \Theta_{HX}\sigma_X L_{LX} + \Theta_{HT}\sigma_T L_{LT} - L_{LR}$$

$$-(1-\Theta_{LX})L_{LX} - (1-\Theta_{LT})L_{LT} + (1-u_L)(1-\phi)\bar{L} - w_L\frac{\partial u_L}{\partial w_L}(1-\phi)\bar{L}\Big]$$

$$+\frac{\breve{w}_H}{(1-\phi)\beta_L\bar{L}}\Big[\Theta_{HR}\sigma_R L_{LR} + \Theta_{HX}\sigma_X L_{LX} + \Theta_{HT}\sigma_T L_{LT} - \Theta_{HX}L_{LX}$$

$$- \Theta_{HT}L_{LT}\Big] + \frac{\iota L_{LR}}{(1-\phi)\beta_L\bar{L}} = -\frac{\partial u_L}{\partial\beta_L}\breve{\beta}_L$$

Noting that $L_{LR} + L_{LX} + L_{LT} = (1-\phi)(1-u_L)\bar{L}$ means that the above equation simplifies to

$$-\frac{\breve{w}_L}{(1-\phi)\beta_L\bar{L}}\Big[\Theta_{HR}\sigma_R L_{LR} + \Theta_{HX}\sigma_X L_{LX} + \Theta_{HT}\sigma_T L_{LT} + \Theta_{LX}L_{LX}$$

$$+\Theta_{LT}L_{LT} - w_L\frac{\partial u_L}{\partial w_L}(1-\phi)\bar{L}\Big] + \frac{\breve{w}_H}{(1-\phi)\beta_L\bar{L}}\Big[\Theta_{HR}\sigma_R L_{LR} + \Theta_{HX}\sigma_X L_{LX}$$

$$+ \Theta_{HT}\sigma_T L_{LT} - \Theta_{HX}L_{LX} - \Theta_{HT}L_{LT}\Big] + \frac{\iota L_{LR}}{(1-\phi)\beta_L\bar{L}} = -\frac{\partial u_L}{\partial\beta_L}\breve{\beta}_L$$

Totally differentiating (6.30) yields:

$$0 = \Big[\frac{\partial c_R}{\partial w_L}\,\mathrm{d}w_L + \frac{\partial c_R}{\partial w_H}\,\mathrm{d}w_H\Big](\rho + \iota(1+\gamma\ln\lambda)) + c_R(1+\gamma\ln\lambda)\,\mathrm{d}\iota$$

$$= (a_{LR}\,\mathrm{d}w_L + a_{HR}\,\mathrm{d}w_H)(\rho + \iota(1+\gamma\ln\lambda)) + c_R(1+\gamma\ln\lambda)\,\mathrm{d}\iota$$

$$= \Theta_{LR}\breve{w}_L + \Theta_{HR}\breve{w}_H + \frac{\iota(1+\gamma\ln\lambda)}{\rho + \iota(1+\gamma\ln\lambda)}\breve{\iota}$$

A.10 Derivation of the Influence of Union Bargaining Power on the Innovation Rate

Using Cramer's rule, the change in the innovation rate associated with an increase in union bargaining strength is

$$\breve{\iota} = \frac{|\mathbf{\Phi}_3|}{|\mathbf{\Phi}|}$$

with $|\mathbf{\Phi}_3|$ as

$$|\boldsymbol{\Phi}_3| = \begin{vmatrix} \varPhi_{11} & \varPhi_{12} & \check{\beta}_L \\ \varPhi_{21} & \varPhi_{22} & 0 \\ \varPhi_{31} & \varPhi_{32} & 0 \end{vmatrix}$$

$$\Rightarrow \check{\iota} = \frac{\varPhi_{32}\check{\beta}_L\varPhi_{21} - \varPhi_{31}\check{\beta}_L\varPhi_{22}}{|\boldsymbol{\Phi}|}$$

$$\frac{\check{\iota}}{\check{\beta}_L} = \frac{1}{\phi\bar{L}|\boldsymbol{\Phi}|}\Bigg[\Theta_{HR}(\Theta_{LR}\sigma_R L_{HR} + \Theta_{LX}\sigma_X L_{HX} + \Theta_{LT}\sigma_T L_{HT} - \Theta_{LX}L_{HX}$$
$$- \Theta_{LT}L_{HT}) + \Theta_{LR}(\Theta_{LR}\sigma_R L_{HR} + \Theta_{LX}\sigma_X L_{HX} + \Theta_{LT}\sigma_T L_{HT}$$
$$+ \Theta_{HX}L_{HX} + \Theta_{HT}L_{HT})\Bigg]$$

References

Abowd, J. M. and Kramarz, F. (2000). Inter-Industry and Firm-Size Wage Differentials: New Evidence from Linked Employer-Employee Data. Cornell University Working Paper.

Acemoglu, D. (1998). Why Do New Technologies Complement Skills? Directed Technical Change and Wage Inequality. *Quarterly Journal of Economics* 113: 1055–1090.

Acemoglu, D. (1999). Changes in Unemployment and Wage Inequality: An Alternative Theory and Some Evidence. *American Economic Review* 89: 1259–1278.

Acemoglu, D. (2001a). Directed Technical Change. NBER Working Paper No. 8287.

Acemoglu, D. (2001b). Good Jobs versus Bad Jobs. *Journal of Labor Economics* 19: 1–21.

Adams, J. S. (1963). Toward an Understanding of Inequity. *Journal of Abnormal and Social Psychology* 67: 422–436.

Addison, J. T. and Grosso, J.-L. (1996). Job Security Provisions and Employment: Revised Estimates. *Industrial Relations* 35: 585–603.

Agell, J. (1999). On the Benefits from Rigid Labour Markets: Norms, Market Failures, and Social Insurance. *Economic Journal* 109: F143–F164.

Agell, J. (2002). On the Determinants of Labour Market Institutions: Rent Seeking vs. Social Insurance. *German Economic Review* 3: 107–135.

Agell, J. and Lommerud, K. E. (1992). Union Egalitarianism as Income Insurance. *Economica* 59: 295–310.

Agell, J. and Lundborg, P. (1992). Fair Wages, Involuntary Unemployment and Tax Policies in the Simple General Equilibrium Model. *Journal of Public Economics* 47: 299–320.

Aghion, P., Caroli, E. and García-Peñalosa, C. (1999). Inequality and Economic Growth: The Perspective of the New Growth Theories. *Journal of Economic Literature* 37: 1615–1660.

Aghion, P. and Howitt, P. (1992). A Model of Growth through Creative Destruction. *Econometrica* 60: 323–351.

Aghion, P. and Howitt, P. (1994). Growth and Unemployment. *Review of Economic Studies* 61: 477–494.

Aghion, P. and Howitt, P. (1998). *Endogenous Growth Theory*. Cambridge, MA and London: MIT Press.

Akerlof, G. A. (1982). Labor Contracts as Partial Gift Exchange. *Quarterly Journal of Economics* 97: 543–569.

Akerlof, G. A. (1984). Gift Exchange and Efficiency-Wage Theory: Four Views. *American Economic Review* 74: 79–83.

Akerlof, G. A. and Yellen, J. L., eds. (1986). *Efficiency Wage Models of the Labor Market*. Cambridge: Cambridge University Press.

Akerlof, G. A. and Yellen, J. L. (1988). Fairness and Unemployment. *American Economic Review* 78: 44–49.

Akerlof, G. A. and Yellen, J. L. (1989). Workers' Trust Funds and the Logic of Wage Profiles. *Quarterly Journal of Economics* 104: 525–536.

Akerlof, G. A. and Yellen, J. L. (1990). The Fair-Wage Effort Hypothesis and Unemployment. *Quarterly Journal of Economics* 105: 255–283.

Albert, M. and Meckl, J. (2001). Efficiency-Wage Unemployment and Intersectoral Wage Differentials in a Heckscher-Ohlin Model. *German Economic Review* 2: 287–301.

Albrecht, J. W. and Vroman, S. B. (2002). A Matching Model with Endogenous Skill Requirements. *International Economic Review* 43: 283–305.

Alchian, A. A. (1969). Information Costs, Pricing, and Resource Unemployment. *Western Economic Journal* 7: 109–128.

Altenburg, L. and Straub, M. (1998). Efficiency Wages, Trade Unions, and Employment. *Oxford Economic Papers* 50: 726–746.

Amable, B. and Gatti, D. (2002). Macroeconomic Effects of Product Market Competition in a Dynamic Efficiency Wage Model. *Economics Letters* 75: 39–46.

Arnsperger, C. and de la Croix, D. (1990). Wage Bargaining with a Price-Setting Firm. *Bulletin of Economic Research* 42: 285–298.

Arrow, K. J. (1950). A Difficulty in the Concept of Social Welfare. *Journal of Political Economy* 58: 328–346.

Arulampalam, W. (2001). Is Unemployment Really Scarring? Effects of Unemployment Experiences on Wages. *Economic Journal* 111: F585–606.

Autor, D., Katz, L. F. and Krueger, A. B. (1998). Computing Inequality: Have Computers Changed the Labor Market. *Quarterly Journal of Economics* 113: 1169–1214.

Azariadis, C. (1975). Implicit Contracts and Underemployment Equilibria. *Journal of Political Economy* 83: 1183–1202.

Baily, M. N. (1974). Wages and Employment under Uncertain Demand. *Review of Economic Studies* 41: 37–50.

Baltagi, B. H. and Blien, U. (1998). The German Wage Curve: Evidence from the IAB Employment Sample. *Economics Letters* 61: 135–142.

Barro, R. J. and Sala-i-Martin, X. (1995). *Economic Growth*. New York et al.: McGraw–Hill.

Barron, J. M. (1975). Search in the Labor Market and the Duration of Unemployment: Some Empirical Evidence. *American Economic Review* 65: 934–942.

Barth, E. and Zweimüller, J. (1995). Relative Wages under Decentralized and Corporatist Bargaining Systems. *Scandinavian Journal of Economics* 97: 369–384.

Bellman, R. (1957). *Dynamic Programming*. Princeton, NJ and Chichester: Princeton University Press.

Bentolila, S. and Bertola, G. (1990). Firing Costs and Labour Demand: How Bad Is Eurosclerosis? *Review of Economic Studies* 57: 381–402.

Berloffa, G. (1997). Temporary and Permanent Changes in Consumption Growth. *Economic Journal* 107: 345–358.

Beveridge, W. H. (1944). *Full Employment in a Free Society*. London: George Allen and Unwin.

Binmore, K., Rubinstein, A. and Wolinsky, A. (1986). The Nash Bargaining Solution in Economic Modelling. *RAND Journal of Economics* 17: 176–188.

Black, D. (1948). On the Rationale of Group Decision Making. *Journal of Political Economy* 56: 23–34.

Blanchard, O. and Giavazzi, F. (2001). The Macroeconomic Effects of Labor and Product Market Deregulation. MIT Working Paper No. 01/02.

Blanchard, O. J. and Diamond, P. A. (1994). Ranking, Unemployment Duration, and Wages. *Review of Economic Studies* 61: 417–434.

Blanchard, O. J. and Summers, L. H. (1986). Hysteresis and the European Unemployment Problem. *NBER Macroeconomics Annual* 1: 15–78.

Blanchflower, D. G. and Oswald, A. J. (1994a). Estimating a Wage Curve for Britain. *Economic Journal* 104: 1025–1043.

Blanchflower, D. G. and Oswald, A. J. (1994b). *The Wage Curve*. Cambridge, MA and London: MIT Press.

Blank, R. M. and Freeman, R. B. (1994). Evaluating the Connection between Social Protection and Economic Flexibility. In: R. M. Blank, ed., *Social Protection versus Economic Flexibility: Is There a Trade-Off?*, NBER Comparative Labor Market Series, Chicago and London: University of Chicago Press, 21–41.

Blau, F. D. and Kahn, L. M. (1996). International Differences in Male Wage Inequality: Institutions versus Market Forces. *Journal of Political Economy* 104: 791–837.

Blundell, R. (1991). Consumer Behaviour: Theory and Empirical Evidence – A Survey. *Surveys in Economics* 2: 1–50.

Booth, A. L. (1984). A Public Choice Model of Trade Union Behaviour. *Economic Journal* 94: 883–898.

Booth, A. L. (1985). The Free Rider Problem and a Social Custom Model of Trade Union Membership. *Quarterly Journal of Economics* 10: 253–261.

Booth, A. L. (1995). *The Economics of the Trade Union*. Cambridge et al.: Cambridge University Press.

Bound, J. and Johnson, G. (1992). Changes in the Structure of Wages in the 1980's: An Evaluation of Alternative Explanations. *American Economic Review* 82: 371–392.

Bound, J. and Johnson, G. (1995). What are the Causes of Rising Wage Inequality in the United States. *Federal Reserve Bank of New York Economic Policy Review* 1: 9–17.

Bräuninger, M. (2000). Wage Bargaining, Unemployment and Growth. *Journal of Institutional and Theoretical Economics* 156: 646–660.

Brown, C. and Medoff, J. (1978). Trade Unions in the Production Process. *Journal of Political Economy* 86: 355–378.

Bulkley, G. and Myles, G. D. (2001). Individually Rational Union Membership. *European Journal of Political Economy* 17: 117–137.

Bulow, J. I. and Summers, L. H. (1986). A Theory of Dual Labor Markets with Application to Industrial Policy, Discrimination, and Keynesian Unemployment. *Journal of Labor Economics* 4: 376–414.

Bundesanstalt für Arbeit (1965–2001). *Amtliche Nachrichten der Bundesanstalt für Arbeit*, Arbeitsstatistik – Jahreszahlen, Nürnberg.

Bundesanstalt für Arbeit (2003). Zahlen-Fibel.
URL: http://iab.de/asp/fibel/default.asp (as of 16..04.03).

Burda, M. (1988). 'Wait Unemployment' in Europe. *Economic Policy* 7: 391–425.

Burda, M. and Wyplosz, C. (1994). Gross Labor Market Flows in Europe: Some Stylized Facts. *European Economic Review* 38: 1287–1325.

Burdett, K. and Mortensen, D. T. (1998). Wage Differentials, Employer Size, and Unemployment. *International Economic Review* 39: 257–273.

Buscher, H., Falk, M., Göggelmann, K., Ludsteck, J., Steiner, V. and Zwick, T. (2000). *Wachstum, Beschäftigung und Arbeitslosigkeit*. ZEW Wirtschaftsanalysen 48, Baden-Baden: Nomos.

Cahuc, P. and Postel-Vinay, F. (2002). Temporary Jobs, Employment Protection and Labor Market Performance. *Labour Economics* 9: 63–91.

Calmfors, L. and Driffill, E. J. (1988). Bargaining Structure, Corporatism and Macroeconomic Performance. *Economic Policy* 6: 14–61.

Calvo, G. A. (1978). Urban Unemployment and Wage Determination in LDC's: Trade Unions in the Harris-Todaro Model. *International Economic Review* 19: 65–81.

Campbell, C. M. (1993). Do Firms Pay Efficiency Wages? Evidence with Data at the Firm Level. *Journal of Labor Economics* 3: 442–470.

Cappelli, P. and Chauvin, K. (1991). An Interplant Test of the Efficiency Wage Hypothesis. *Quarterly Journal of Economics* 106: 769–787.

Card, D. (1995). The Wage Curve: A Review. *Journal of Economic Literature* 33: 785–799.

Carmichael, L. (1985). Can Unemployment Be Involuntary?: Comment. *American Economic Review* 75: 1213–1214.

Caroli, E. and van Reenen, J. (2001). Skill-Biased Organizational Change? Evidence from a Panel of British and French Establishments. *Quarterly Journal of Economics* 116: 1448–1492.

Caselli, F. (1999). Technological Revolutions. *American Economic Review* 89: 78–102.

Cass, D. (1965). Optimum Growth in an Aggregative Model of Capital Accumulation. *Review of Economic Studies* 32: 233–240.

Chamberlin, E. H. (1933). *The Theory of Monopolistic Competition*. Harvard Economic Studies 38, Cambridge, MA: Harvard University Press.

Coles, M. G. (1999). Turnover Externalities with Marketplace Trading. *International Economic Review* 40: 851–868.

Coles, M. G. and Muthoo, A. (1998). Strategic Bargaining and Competitive Bidding in a Dynamic Market Equilibrium. *Review of Economic Studies* 65: 235–260.

Coles, M. G. and Smith, E. (1998). Marketplaces and Matching. *International Economic Review* 39: 239–254.

Daveri, F. and Tabellini, G. (2000). Unemployment, Growth and Taxation in Industrial Countries. *Economic Policy* 15: 47–104.

Davidson, C., Martin, L. and Matusz, S. (1988). The Structure of Simple General Equilibrium Models with Frictional Unemployment. *Journal of Political Economy* 96: 1267–1293.

Davis, S. J. and Haltiwanger, J. (1990). Job Creation, Job Destruction, and Job Reallocation over the Cycle. In: O. J. Blanchard and S. Fisher, eds., *NBER Macroeconomics Annual 5*, Cambridge, MA and London: MIT Press, 123–168.

Davis, S. J. and Haltiwanger, J. (1992). Gross Job Creation, Gross Job Destruction, and Employment Reallocation. *Quarterly Journal of Economics* 107: 819–864.

Davis, S. J. and Haltiwanger, J. (1999). Gross Job Flows. In: O. Ashenfelter and R. Layard, eds., *Handbook of Labor Economics*, Vol. 3B, Amsterdam et al.: North–Holland, 2711–2805.

de Groot, H. L. (2000). *Growth, Unemployment and Deindustrialization*. Cheltenham and Northampton, MA: Edward Elgar.

Deardorff, A. V. and Hakura, D. S. (1994). Trade and Wages: What Are the Questions? In: J. N. Bhagwati and M. H. Kosters, eds., *Trade and Wages: Leveling Wages Down*, Washington D.C.: AEI Press, 76–107.

Desjonqueres, T., Machin, S. and van Reenen, J. (1999). Another Nail in the Coffin? Or Can the Trade Based Explanation of Changing Skill Structures Be Resurrected? *Scandinavian Journal of Economics* 101: 533–554.

Deutsche Bundesbank (1995). Das Produktionspotential in Deutschland und seine Bestimmungsfaktoren. *Monatsberichte* 47: 41–56.

Deutscher Gewerkschaftsbund (2002). Mitglieder in den DGB-Gewerkschaften 2000. URL: http://www.dgb.de/wir/statistik/statistik.index.html (as of 10.06.02).

Diamond, P. A. (1982). Wage Determination and Efficiency in Search Equilibrium. *Review of Economic Studies* 49: 217–227.

Dickens, W. T. and Katz, L. F. (1987). Inter-industry Wage Differences and Industry Characteristics. In: K. Lang and J. S. Leonard, eds., *Unemployment and the Structure of Labor Markets*, Oxford and Cambridge, MA: Blackwell, 48–89.

Dickens, W. T. and Lang, K. (1993). Labor Market Segmentation Theory: Reconsidering the Evidence. In: W. A. Darity, ed., *Labor Economics: Problems in Analyzing Labor Markets*, Boston and London: Kluwer Academic Publishers, 141–180.

Dinopoulos, E. and Thompson, P. (1998). Schumpeterian Growth Without Scale Effects. *Journal of Economic Growth* 3: 313–335.

Dinopoulos, E. and Thompson, P. (1999). Scale Effects in Schumpeterian Models of Economic Growth. *Journal of Evolutionary Economics* 9: 157–185.

Dixit, A. K. and Stiglitz, J. E. (1977). Monopolistic Competition and Optimum Product Diversity. *American Economic Review* 67: 297–308.

Dixon, H. D., Hansen, C. T. and Kleven, H. J. (1999). Dual Labour Markets and Menu Costs: Explaining the Cyclicality of Productivity and Wage Differentials. University of York Discussion Paper No.99-01.

Dolado, J., Kramarz, F., Machin, S., Manning, A., Margolis, D. and Teulings, C. (1996). The Economic Impact of Minimum Wages in Europe. *Economic Policy* 11: 317–372.

Dunlop, J. T. (1944). *Wage Determination under Trade Unions*. New York et al.: Macmillan.

Dutt, A. K. and Sen, A. (1997). Union Bargaining Power, Employment and Output in a Model of Monopolistic Competition with Wage Bargaining. *Journal of Economics* 65: 1–17.

Eguchi, K. (2002). Unions as Commitment Devices. *Journal of Economic Behavior & Organization* 47: 407–421.

Eichengreen, B. and Iversen, T. (1999). Institutions and Economic Performance: Evidence from the Labour Market. *Oxford Review of Economic Policy* 15: 121–138.

Eicher, T. S. (1996). Interaction Between Endogenous Human Capital and Technological Change. *Review of Economic Studies* 63: 127–144.

Elmeskov, J., Martin, J. P. and Scarpetta, S. (1998). Key Lessons for Labour Market Reforms: Evidence from OECD Countries' Experiences. *Swedish Economic Policy Review* 5: 205–252.

Entorf, H. (1996). Strukturelle Arbeitslosigkeit in Deutschland: Mismatch, Mobilität und technischer Wandel. In: B. Gahlen, H. Hesse and H. J. Ramser, eds., *Arbeitslosigkeit und Möglichkeiten ihrer Überwindung*, Wirtschaftswissenschaftliches Seminar Ottobeuren 25, Tübingen: Mohr Siebeck, 139–170.

Eriksson, C. (1997). Is There a Trade-Off between Employment and Growth? *Oxford Economic Papers* 49: 77–88.

Ewing, B. T. and Payne, J. E. (1999). The Trade-Off between Supervision and Wages: Evidence of Efficiency Wages from the NLSY. *Southern Economic Journal* 66: 424–432.

Farber, H. S. (1986). The Analysis of Union Behavior. In: O. Ashenfelter and R. Layard, eds., *Handbook of Labor Economics*, Vol. 2, Amsterdam et al.: North–Holland, 1039–1089.

Fitzenberger, B. (1999). *Wages and Employment Across Skill Groups. An Analysis for West Germany.* ZEW Economic Studies, 6, Heidelberg: Physica-Verlag.

Fitzenberger, B. and Franz, W. (1998). Flexibilität der qualifikatorischen Lohnstruktur und Lastverteilung der Arbeitslosigkeit: Eine ökonometrische Analyse für Westdeutschland. In: B. Gahlen, H. Hesse and H. J. Ramser, eds., *Verteilungsprobleme der Gegenwart – Diagnose und Therapie*, Wirtschaftswissenschaftliches Seminar Ottobeuren 27, Tübingen: Mohr Siebeck, 47–79.

Flaig, G., Licht, G. and Steiner, V. (1993). Testing for State Dependence Effects in a Dynamic Model of Male Unemployment Behaviour. In: H. Bunzel, P. Jensen and N. Westergård-Nielsen, eds., *Panel Data and Labour Market Dynamics*, Amsterdam et al.: North–Holland, 189–213.

Flaig, G. and Rottmann, H. (2001). Input Demand and the Short- and Long-Run Employment Thresholds. An Empirical Analysis for the German Manufacturing Sector. *German Economic Review* 2: 367–384.

Fonseca, R., Lopez-Garcia, P. and Pissarides, C. A. (2002). Entrepreneurship, Start-Up Costs and Employment. *European Economic Review* 45: 692–705.

Franz, W. (2003). *Arbeitsmarktökonomik*. 5th rev. ed., Berlin et al.: Springer.

Fredriksson, P. and Holmlund, B. (2001). Optimal Unemployment Insurance in Search Equilibrium. *Journal of Labor Economics* 19: 370–399.

Gächter, S. and Falk, A. (2002). Reputation and Reciprocity: Consequences for the Labour Relation. *Scandinavian Journal of Economics* 104: 1–26.

Garibaldi, P. (1998). Job Flow Dynamics and Firing Restrictions. *European Economic Review* 42: 245–275.

Gautier, P. A. (2002). Unemployment and Search Externalities in a Model with Heterogeneous Jobs and Workers. *Economica* 69: 21–40.

Gautier, P. A., van den Berg, G. J., van Ours, J. C. and Ridder, G. (1999). Separations at the Firm Level. In: J. C. Haltiwanger, J. I. Lane, J. R. Spetzler, J. Theeuwes and K. R. Troske, eds., *The Creation and Analysis of Matched Employer-Employee Data*, Amsterdam et al.: Elsevier, 313–327.

Gautier, P. A., van den Berg, G. J., van Ours, J. C. and Ridder, G. (2002). Worker Turnover at the Firm Level and Crowding Out of Lower Educated Workers. *European Economic Review* 46: 523–538.

Gera, S. and Grenier, G. (1994). Interindustry Wage Differentials and Efficiency Wages: Some Canadian Evidence. *Canadian Journal of Economics* 27: 81–100.

Gibbons, R. and Katz, L. F. (1992). Does Unmeasured Ability Explain Inter-Industry Wage Differentials. *Review of Economic Studies* 59: 515–535.

Goerke, L. (2000). On the Structure of Unemployment Benefits in Shirking Models. *Labour Economics* 7: 283–295.

Goerke, L. and Holler, M. J. (1997). *Arbeitsmarktmodelle*. Berlin et al.: Springer.

Gordon, D. F. (1974). A Neo-Classical Theory of Keynesian Unemployment. *Economic Inquiry* 12: 431–459.

Gregg, P. (2001). The Impact of Youth Unemployment on Adult Unemployment in the NCDS. *Economic Journal* 111: F626–F653.

Gregg, P. and Manning, A. (1997). Skill-Biased Change, Unemployment and Wage Inequality. *European Economic Review* 41: 1173–1200.

Gregory, M. and Jukes, R. (2001). Unemployment and Subsequent Earnings: Estimating Scarring among British Men 1984–94. *Economic Journal* 111: F607–F625.

Grossman, G. M. and Helpman, E. (1991a). *Innovation and Growth in the Global Economy*. Cambridge, MA and London: MIT Press.

Grossman, G. M. and Helpman, E. (1991b). Quality Ladders in the Theory of Growth. *Review of Economic Studies* 58: 43–61.

Grossman, V. (2000). Skilled Labor Reallocation, Wage Inequality, and Unskilled Unemployment. *Journal of Institutional and Theoretical Economics* 156: 473–500.

Gruber, J. (2001). The Wealth of the Unemployed. *Industrial and Labor Relations Review* 55: 79–94.

Haisken-DeNew, J. P. and Schmidt, C. M. (1999). Industry Wage Differentials Revisited: A Longitudinal Comparison of Germany and USA 1984–1996. IZA Discussion Paper No. 98.

Hamermesh, D. S. (1986). The Demand for Labor in the Long Run. In: O. Ashenfelter and R. Layard, eds., *Handbook of Labor Economics*, Vol. 1, Amsterdam et al.: North–Holland, 429–471.

Hamermesh, D. S. (1993). *Labor Demand*. Princeton, NJ and Chichester: Princeton University Press.

Hansen, G. D. and İmrohoroğlu, A. (1992). The Role of Unemployment Insurance in an Economy with Liquidity Constraints and Moral Hazard. *Journal of Political Economy* 100: 118–142.

Harris, J. R. and Todaro, M. P. (1970). Migration, Unemployment and Development: A Two-Sector Analysis. *American Economic Review* 60: 126–142.

Hart, O. (1982). A Model of Imperfect Competition with Keynesian Features. *Quarterly Journal of Economics* 97: 109–138.

Haskel, J. E. and Slaughter, M. J. (2002). Does the Sector Bias of Skill-Biased Technical Change Explain Changing Skill Premia? *European Economic Review* 46: 1757–1783.

Holzer, H. J., Katz, L. F. and Krueger, A. B. (1991). Job Queues and Wages. *Quarterly Journal of Economics* 106: 739–768.

Inada, K.-I. (1963). On a Two-Sector Model of Economic Growth. *Review of Economic Studies* 30: 119–127.

Irmen, A. and Wigger, B. U. (2001). Trade Union Objectives and Economic Growth. CEPR Discussion Paper No. 3027.

Jean, S. and Nicoletti, G. (2002). Product Market Regulation and Wage Premia in Europe and North America: An Empirical Investigation. OECD Economics Department Working Paper No. 318.

Johnson, G. E. (1997). Changes in Earnings Inequality. The Role of Demand Shifts. *Journal of Economic Perspectives* 11: 41–54.

Jones, R. W. (1965). The Structure of Simple General Equilibrium Models. *Journal of Political Economy* 73: 557–572.

Jones, S. R. (1987a). Minimum Wage Legislation in a Dual Labor Market. *European Economic Review* 31: 1229–1246.

Jones, S. R. (1987b). Screening Unemployment in a Dual Labor Market. *Economics Letters* 25: 191–195.

Juhn, C., Murphy, K. M. and Pierce, B. (1993). Wage Inequality and the Rise in Returns to Skill. *Journal of Political Economy* 101: 410–442.

Kahn, L. M. (1998). Collective Wage Bargaining and the Interindustry Wage Structure: International Evidence. *Economica* 65: 507–534.

Kahneman, D., Knetsch, J. L. and Thaler, R. (1986). Fairness as a Constraint on Profit Seeking: Entitlements in the Market. *American Economic Review* 76: 728–741.

Katz, L. F. (1986). Efficiency Wage Theories: A Partial Evaluation. In: S. Fischer, ed., *NBER Macroeconomics Annual* 1, Cambridge, MA and London: MIT Press, 235–276.

Katz, L. F. (2000). Technological Change, Computerization, and the Wage Structure. In: E. Brynjolfsson and B. Kahin, eds., *Understanding the Digital Economy*, Cambridge, MA and London: MIT Press, 217–244.

Katz, L. F. and Murphy, K. M. (1992). Changes in Relative Wages, 1963 – 1987. Supply and Demand Factors. *Quarterly Journal of Economics* 107: 35–78.

Katz, L. F. and Summers, L. H. (1989). Industry Rents: Evidence and Implications. *Brookings Papers on Economic Activity:* Microeconomics: 209–275.

Kennan, J. (1986). The Economics of Strikes. In: O. Ashenfelter and R. Layard, eds., *Handbook of Labor Economics*, Vol. 2, Amsterdam et al.: North–Holland, 1091–1137.

Kiley, M. T. (1999). The Supply of Skilled Labour and Skill-Biased Technological Progress. *Economic Journal* 109: 708–724.

Kletzer, L. G. (1992). Industry Wage Differentials and Wait Unemployment. *Journal of Economic Perspectives* 12: 115–136.

Kohns, S. (2000). Different Skill Levels and Firing Costs in a Matching Model with Uncertainty. An Extension of Mortensen and Pissarides (1994). IZA Working Paper No. 104.

Koopmans, T. C. (1965). On the Concept of Optimal Economic Growth. In: *The Econometric Approach to Development Planning*, Pontificiae Academiae Scientiarum Scripta Varia 28, Amsterdam and Chicago: North–Holland, 225–287.

Kremer, M. and Maskin, E. (1996). Wage Inequality and Segregation by Skill. NBER Working Paper No. 5718.

Krueger, A. B. and Summers, L. H. (1988). Efficiency Wages and the Inter-Industry Wage Structure. *Econometrica* 56: 259–293.

Krugman, P. (1994). Past and Prospective Causes of High Unemployment. In: *Reducing Unemployment: Current Issues and Policy Options.* A Symposium Sponsored by the Federal Reserve Bank of Kansas City, Jackson Hole, Wyoming, August 25–27, 1994, 49–80.

Laing, D. (1993). A Signalling Theory of Nominal Wage Inflexibility. *Economic Journal* 103: 1493–1510.

Layard, R. and Nickell, S. (1990). Is Unemployment Lower if Unions Bargain over Unemployment? *Quarterly Journal of Economics* 105: 773–787.

Layard, R., Nickell, S. and Jackman, R. (1991). *Unemployment: Macroeconomic Performance and the Labour Market.* Oxford et al.: Oxford University Press.

Lazear, E. P. (1990). Job Security Provisions and Employment. *Quarterly Journal of Economics* 105: 699–726.

Leibenstein, H. (1957). *Economic Backwardness and Economic Growth: Studies in the Theory of Economic Development.* New York and London: John Wiley & Sons.

Lewis, H. G. (1963). *Unionism and Relative Wages in the United States.* Chicago and London: University of Chicago Press.

Lewis, H. G. (1986). *Union Relative Wage Effects: A Survey.* Chicago and London: University of Chicago Press.

Li, C.-W. (2000). Endogenous vs. Semi-Endogenous Growth in a Two-Sector R&D Model. *Economic Journal* 110: C109–C122.

Li, C.-W. (2001). On the Policy Implications of Endogenous Technological Progress. *Economic Journal* 111: C164–C179.

Li, C.-W. (2002). Growth and Scale Effects: The Role of Knowledge Spillovers. *Economics Letters* 74: 177–185.

Lindbeck, A. and Snower, D. J. (1989). *The Insider-Outsider Theory of Employment and Unemployment.* Cambridge, MA and London: MIT Press.

Lindbeck, A. and Snower, D. J. (1996). Reorganization of Firms and Labor-Market Inequality. *American Economic Review* 86: 315–321.

Lindbeck, A. and Snower, D. J. (2001). Insiders versus Outsiders. *Journal of Economic Perspectives* 15: 165–188.

Lucas, R. E., Jr. (1988). On the Mechanics of Economic Development. *Journal of Monetary Economics* 22: 3–42.

Ma, C. and Weiss, A. M. (1993). A Signaling Theory of Unemployment. *European Economic Review* 37: 135–157.

Machin, S. and van Reenen, J. (1998). Technology and Changes in Skill Structure: Evidence from Seven OECD Countries. *The Quarterly Journal of Economics* 113: 1215–1244.

McConnell, C. R., Brue, S. L. and Macpherson, D. A. (1999). *Contemporary Labor Economics.* 5th ed., New York et al.: McGraw–Hill.

McCormick, B. (1990). A Theory of Signalling during Job Search, Employment Efficiency, and "Stigmatised" Jobs. *Review of Economic Studies* 57: 299–313.

McDonald, I. M. and Solow, R. M. (1981). Wage Bargaining and Employment. *American Economic Review* 71: 896–908.

McDonald, I. M. and Solow, R. M. (1985). Wages and Employment in a Segmented Labor Market. *Quarterly Journal of Economics* 100: 1115–1141.

McKenna, C. J. (1996). Education and the Distribution of Unemployment. *European Journal of Political Economy* 12: 113–132.

Mincer, J. (1995). Economic Development, Growth of Human Capital, and the Dynamics of the Wage Structure. *Journal of Economic Growth* 1: 29–48.

Moen, E. R. (1997). Competitive Search Equilibrium. *Journal of Political Economy* 105: 385–411.

Mortensen, D. T. (1986). Job Search and Labor Market Analysis. In: O. Ashenfelter and R. Layard, eds., *Handbook of Labor Economics*, Vol. 2, Amsterdam et al.: North–Holland, 849–919.

Mortensen, D. T. and Pissarides, C. A. (1994). Job Creation and Job Destruction in the Theory of Unemployment. *Review of Economic Studies* 61: 397–415.

Mortensen, D. T. and Pissarides, C. A. (1999a). Job Reallocation, Employment Fluctuations and Unemployment. In: J. B. Taylor and M. Woodford, eds., *Handbook of Macroeconomics*, Vol. 1B, Amsterdam et al.: North–Holland, 1171–1228.

Mortensen, D. T. and Pissarides, C. A. (1999b). New Developments in Models of Search in the Labor Market. In: O. Ashenfelter and D. Card, eds., *Handbook of Labor Economics*, Vol. 3B, Amsterdam et al.: North–Holland, 2567–2627.

Mortensen, D. T. and Pissarides, C. A. (1999c). Unemployment Responses to "Skill-Biased" Technology Shocks: The Role of Labour Market Policy. *Economic Journal* 109: 242–265.

Muhleisen, M. and Zimmermann, K. (1994). A Panel Analysis of Job Changes and Unemployment. *European Economic Review* 38: 793–801.

Nash, J. (1950). The Bargaining Problem. *Econometrica* 18: 155–162.

Nickell, S. and Bell, B. (1995). The Collapse in Demand for the Unskilled and Unemployment Across the OECD. *Oxford Review of Economic Policy* 11: 40–62.

Nickell, S. and Bell, B. (1996). Changes in the Distribution of Wages and Unemployment in OECD Countries. *American Economic Review (Papers and Proceedings)* 86: 302–308.

Nickell, S. and Layard, R. (1999). Labor Market Institutions and Economic Performance. In: O. Ashenfelter and D. Card, eds., *Handbook of Labor Economics*, Vol. 3C, Amsterdam et al.: North–Holland, 3029–3084.

Nicoletti, G., Bassanini, A., Ernst, E., Jean, S., Santiago, P. and Swaim, P. (2002). Product and Labour Markets Interactions in OECD Countries. OECD Economics Department Working Paper No. 312.

OECD (1965–2001). *Quarterly Labour Force Statistics.*

OECD (1987–2002). *Economic Outlook*, Vol. 42–71.

OECD (1993). Employment Outlook. Paris.

OECD (1994a). *Employment Outlook.*

OECD (1994b). *The OECD Jobs Study; Evidence and Explanations, Part II: The Adjustment Potential of the Labour Market.*

OECD (1995-2001). *Education at a Glance – OECD Indicators.*

OECD (1996). *Employment Outlook.*

OECD (1997). *Employment Outlook.*

OECD (1997–2002). *Employment Outlooks.*

OECD (1998). *Fostering Entrepreneurship.*

OECD (1999). *Economic Outlook*, Vol. 65.

OECD (2000). *Science, Technology and Industry Outlook.*

OECD (2001a). *Economic Outlook*, Vol. 69.

OECD (2001b). *Employment Outlook.*

OECD (2001c). *Labour Force Statistics 1980 – 2000.*

OECD (2002). *Employment Outlook.*

Olson, M. (1984). *The Rise and Decline of Nations: Economic Growth, Stagflation, and Social Rigidities.* New Haven and London: Yale University Press.

Oswald, A. J. (1982). The Microeconomic Theory of the Trade Union. *The Economic Journal* 92: 576–595.

Oswald, A. J. (1985). The Economic Theory of Trade Unions: An Introductory Survey. *Scandinavian Journal of Economics* 87: 160–193.

Oswald, A. J. (1993). Efficient Contracts Are on the Labour Demand Curve: Theory and Facts. *Labour Economics* 1: 85–113.

Palokangas, T. (1996). Endogenous Growth and Collective Bargaining. *Journal of Economic Dynamics & Control* 20: 925–944.

Pencavel, J. H. (1994). *Labor Markets under Trade Unionism: Employment, Wages, and Hours.* Oxford and Cambridge, MA: Blackwell.

Petrongolo, B. and Pissarides, C. A. (2001). Looking into the Black Box: A Survey of the Matching Function. *Journal of Economic Literature* 39: 390–431.

Phelps, E. S. (1968). Money-Wage Dynamics and Labor-Market Equilibrium. *Journal of Political Economy* 76: 678–711.

Phelps, E. S. (1994). *Structural Slumps: The Modern Equilibrium Theory of Unemployment, Interest and Assets.* Cambridge, MA: Harvard University Press.

Phelps, E. S. (1995). The Structuralist Theory of Employment. *American Economic Review (Papers and Proceedings)* 85: 226–231.

Pichler, E. (1993). Efficiency Wages and Union Wage Bargaining. *Jahrbücher für Nationalökonomie und Statistik* 212: 140–150.

Pissarides, C. A. (1979). Job Matchings with State Employment Agencies and Random Search. *Economic Journal* 89: 818–833.

Pissarides, C. A. (1985). Short-Run Equilibrium Dynamics of Unemployment, Vacancies, and Real Wages. *American Economic Review* 75: 676–690.

Pissarides, C. A. (1998). The Impact of Employment Tax Cuts on Unemployment and Wages; The Role of Unemployment Benefits and Tax Structure. *European Economic Review* 42: 155–183.

Pissarides, C. A. (2000). *Equilibrium Unemployment Theory*. 2nd ed., Cambridge, MA and London: MIT Press.

Raff, D. M. G. and Summers, L. H. (1987). Did Henry Ford Pay Efficiency Wages? *Journal of Labor Economics* 5: S57–S86.

Ramser, H. J. (1997). Beschäftigung und Wirtschaftswachstum. *ifo Studien* 43: 347–365.

Ramsey, F. (1928). A Mathematical Theory of Saving. *Economic Journal* 38: 543–559.

Rebitzer, J. B. (1995). Is There a Trade-Off between Supervision and Wages? An Empirical Test of Efficiency Wage Theory. *Journal of Economic Behavior and Organization* 28: 107–129.

Reinberg, A. (1999). Der qualifikatorische Strukturwandel auf dem deutschen Arbeitsmarkt – Entwicklungen, Perspektiven und Bestimmungsgründe. *Mitteilungen aus der Arbeitsmarkt und Berufsforschung* 32: 434–447.

Roberts, M. A., Stæhr, K. and Tranæs, T. (2000). Two-Stage Bargaining with Coverage Extension in a Dual Labour Market. *European Economic Review* 44: 181–200.

Romer, P. M. (1986). Increasing Returns and Long Run Growth. *Journal of Political Economy* 94: 1002–1037.

Rosen, S. (1985). Implicit Contracts: A Survey. *Journal of Economic Literature* 23: 1144–1175.

Ross, A. M. (1944). *Trade Union Wage Policy*. Berkeley and Los Angeles: University of California Press.

Rothschild, M. (1973). Models of Market Organization with Imperfect Information: A Survey. *Journal of Political Economy* 81: 1283–1308.

Rubinstein, A. (1982). Perfect Equilibrium in a Bargaining Model. *Econometrica* 50: 97–109.

Saint-Paul, G. (1996a). Are the Unemployed Unemployable? *European Economic Review* 40: 1501–1519.

Saint-Paul, G. (1996b). *Dual Labor Markets: A Macroeconomic Perspective*. Cambridge, MA and London: MIT Press.

Salop, S. C. (1979). A Model of the Natural Rate of Unemployment. *American Economic Review* 69: 117–125.

Sapsford, D. and Tzannatos, Z. (1993). *The Economics of the Labour Market*. Basingstoke et al.: Macmillan.

Schlicht, E. (1978). Labour Turnover, Wage Structure and Natural Unemployment. *Zeitschrift für die gesamte Staatswissenschaft* 134: 337–346.

Schumpeter, J. A. (1942). *Capitalism, Socialism and Democracy*. New York: Harper & Brothera Publishers.

Şener, M. F. (2001). Schumpeterian Unemployment, Trade and Wages. *Journal of International Economics* 54: 119–148.

Shapiro, C. and Stiglitz, J. E. (1984). Involuntary Unemployment as a Worker Discipline Device. *American Economic Review* 74: 433–444.

Shapiro, C. and Stiglitz, J. E. (1985). Can Unemployment Be Involuntary?: Reply. *American Economic Review* 75: 1215–1217.

Solow, R. M. (1956). A Contribution to the Theory of Economic Growth. *Quarterly Journal of Economics* 70: 65–94.

Solow, R. M. (1979). Another Possible Source of Wage Rigidity. *Journal of Macroeconomics* 1: 79–82.

Solow, R. M. (1986). Unemployment: Getting the Question Right. *Economica* 53: S23–S34.

Spence, M. A. (1973). Job Market Signaling. *Quarterly Journal of Economics* 87: 355–374.

Spence, M. A. (1976). Product Selection, Fixed Costs, and Monopolistic Competition. *Review of Economic Studies* 43: 217–235.

Stadler, M. (1996). Elemente und Funktionsweise des strukturalistischen Ansatzes zur Erklärung der Arbeitlosigkeit. In: B. Gahlen, H. Hesse and H. J. Ramser, eds., *Arbeitslosigkeit und Möglichkeiten ihrer Überwindung*, Wirtschaftswissenschaftliches Seminar Ottobeuren 25, Tübingen: Mohr Siebeck, 307–326.

Stadler, M. (1999). Dual Labor Markets, Unemployment and Endogenous Growth. *ifo Studien* 45: 283–301.

Stadler, M. (2001). Demand-Pull and Technology-Push Effects in the Quality-Ladder Model. In: S. K. Berninghaus and M. Braulke, eds., *Beiträge zur Mikro- und Makroökonomik – Festschrift für Hans Jürgen Ramser*, Berlin et al.: Springer, 449–460.

Stadler, M. and Wapler, R. (2001). Endogenous Skilled-Biased Technological Change and Matching Unemployment. Paper presented at the 14th Annual Conference of the European Association of Labour Economists (EALE) from 19th – 22nd September 2002 in Paris. University of Tübingen Discussion Paper No. 220.

Steiner, V. and Mohr, R. (1998). Industrial Change, Stability of Relative Earnings, and Substitution of Unskilled Labor in West Germany. ZEW Discussion Paper No. 98-22.

Steiner, V. and Wagner, K. (1998). Relative Earnings and the Demand for Unskilled Labor in West German Manufacturing. In: S. W. Black, ed., *Globalization, Technological Change, and Labor Markets*, Boston and London: Kluwer Academic Publishers, 89–111.

Stiglitz, J. E. (1986). Theories of Wage Rigidity. In: J. L. Butkiewicz, K. J. Koford and J. B. Miller, eds., *Keynes' Economic Legacy: Contemporary Economic Theories*, New York et al.: Praeger Publishers, 153–206.

Summers, L. H. (1988). Relative Wages, Efficiency Wages, and Keynesian Unemployment. *American Economic Review (Papers and Proceedings)* 78: 383–388.

Sutton, J. (1986). Non-Cooperative Bargaining Theory: An Introduction. *Review of Economic Studies* 53: 709–724.

Swan, T. W. (1956). Economic Growth and Capital Accumulation. *Economic Record* 32: 334–361.

Taylor, C. R. (1995). The Long Side of the Market and the Short End of the Stick: Bargaining Power and Price Formation in Buyers', Sellers', and Balanced Markets. *Quarterly Journal of Economics* 110: 837–855.

Tirole, J. (1988). *The Theory of Industrial Organization.* Cambridge, MA and London: MIT Press.

Tobin, J. (1972). Inflation and Unemployment. *American Economic Review* 62: 1–18.

van Schaik, A. B. and de Groot, H. L. (1998). Unemployment and Endogenous Growth. *Labour* 12: 189–219.

Varian, H. R. (1992). *Microeconomic Analysis.* 3rd ed., New York et al.: Norton.

Wapler, R. (1999). Dual Labour Markets. A Survey. University of Tübingen Discussion Paper No. 166.

Wapler, R. (2001). Unions, Efficiency Wages, and Unemployment. University of Tübingen Discussion Paper No. 210.

Weiss, A. M. (1980). Job Queues and Layoffs in Labor Markets with Flexible Wages. *Journal of Political Economy* 88: 526–538.

Wood, A. (1994). *North-South Trade, Employment and Inequality: Changing Fortunes in a Skill-Driven World.* Oxford: Clarendon Press.

Wood, A. (1998). Globalisation and the Rise in Labour Market Inequalities. *Economic Journal* 108: 1463–1482.

Yellen, J. L. (1984). Efficiency Wage Models of Unemployment. *American Economic Review (Papers and Proceedings)* 74: 200–205.

Young, A. (1998). Growth Without Scale Effects. *Journal of Political Economy* 106: 41–63.

Index

Lecture Notes in Economics and Mathematical Systems

For information about Vols. 1–434
please contact your bookseller or Springer-Verlag